MORPHOMETRICS IN EVOLUTIONARY BIOLOGY

*The Geometry of Size and Shape Change,
With Examples from Fishes*

Fred L. Bookstein
University of Michigan

Barry Chernoff
Academy of Natural Sciences of Philadelphia

Ruth L. Elder
Oberlin College

Julian M. Humphries, Jr.
University of Michigan

Gerald R. Smith
University of Michigan

Richard E. Strauss
University of Michigan

SPECIAL PUBLICATION 15
THE ACADEMY OF NATURAL SCIENCES OF PHILADELPHIA
1985

Library of Congress Cataloging-in-Publication Data
Main entry under title:

Morphometrics in evolutionary biology.

  (Special publication no. 15)
  Bibliography: p
  Includes index.
  1. Morphology—Statistical methods. 2. Morphology—
Mathematics. 3. Evolution. 4. Numerical taxonomy.
5. Fishes—Morphology. 5. Fishes—Evolution.
7. Fishes—Classification. I. Bookstein, Fred L.,
1947-   . II. Series: Special publication
(Academy of Natural Sciences of Philadelphia) ; no. 15.
QH351.M62   1985   597'.04   85-15038

Special Publication 15, The Academy of Natural Sciences of Philadelphia

Copyright © 1985, The Academy of Natural Sciences of Philadelphia

ISBN 0-910006-47-4 Hardcover
ISBN 0-910006-48-2 Softcover

Library of Congress Catalog Card Number 85-71646

Typeset by Delmas Typesetting, Ann Arbor, Michigan
Printed by Braun-Brumfield Inc., Ann Arbor, Michigan

Table of Contents

Preface xi
List of Tables ix
List of Figures xiii

1 Introduction 1
   1.1 Morphometrics in the Context of Systematics 1
   1.2 Selection of Models and Methods 2
       1.2.1 Homology and homologous points 4
       1.2.2 Computing variables from comparisons 9
       1.2.3 Two basic models for morphometrics 6
           1.2.3.1 Form-change as deformation 10
           1.2.3.2 Size and shape as factors 11

2 Foundations 13
   2.1 Rates of Change at a Point 14
       2.1.1 General propositions 14
       2.1.2 Shape change as a relation of coordinate systems 14
           2.1.2.1 The Cartesian grid 14
           2.1.2.2 Principal axes computed from coordinate mesh 16
       2.1.3 Shape change as a deformation of points considered in one
             coordinate system 19
           2.1.3.1 Line-elements in a triangle 20
           2.1.3.2 Principal axes computed from triangles of landmarks .. 22
   2.2 The Derived Data of Size and Shape 23
       2.2.1 Logarithmic transformations 23
           2.2.1.1 The use of covariance matrices 26
       2.2.2 Two comments on size adjustment 26
           2.2.2.1 Why size adjustment cannot adjust for size 27

Contents

    2.2.2.2 The Kluge-Kerfoot "phenomenon": artifact of size correction 35

  2.3 Why Not to Omit Homology Information 37
    2.3.1 Forms as geometrical functions 37
    2.3.2 The confounding of homology in purely geometrical analysis: Examples 40

3 Data Collection and Preparation 45

  3.1 Patterns of Distance Measurements 45
    3.1.1 Conventional distance data sets 45
    3.1.2 Triangulations 46
    3.1.3 Coordinate data 50

  3.2 The Truss Network 52
    3.2.1 Archiving the form with redundancy 52
    3.2.2 Adjustment for measurement error: flattening the truss ... 54
    3.2.3 Examples 57

  3.3 Digitizing 61
    3.3.1 Forms without landmarks 63
    3.3.2 Forms with landmarks 64
    3.3.3 Example: Atherinids 66

4 Analytic Methods 69

  4.1 The Medial Axis 70
    4.1.1 Principles 71
    4.1.2 Example: Opercles 73

  4.2 Path Analysis in Multivariate Morphometrics 78
    4.2.1 Regression models and the consistency criterion 79
    4.2.2 Discriminant analysis 82
    4.2.3 Factor analysis 85
      4.2.3.1 Size as an unmeasured variable 85
      4.2.3.2 How factors impute correlations 87
      4.2.3.3 Comparison with principal components analysis: Wright's Leghorn data 89
      4.2.3.4 Data with a geometric structure: two examples .. 94
    4.2.4 Concluding remarks 101

Contents

4.3 The Shear: Size-Free Discrimination 101
    4.3.1 Discrimination as a component analysis 102
    4.3.2 Calculation of a sheared component 105
    4.3.3 The role of secondary factors in discrimination 110
    4.3.4 Use of the shear method to identify hybrids and unknowns 115
        4.3.4.1 Intermediacy due to hybridization 115
        4.3.4.2 A model for hybrid recognition 118
        4.3.4.3 Identification of unknowns 121
    4.3.5 Shearing in the presence of secondary factors 122

4.4 Computing Average Forms from Truss Networks 124

4.5 The Method of Biorthogonal Grids 127
    4.5.1 Introduction 127
    4.5.2 The biorthogonal method 127
        4.5.2.1 The deformation 127
        4.5.2.2 The biorthogonal coordinate system 128
    4.5.3 Examples 131
        4.5.3.1 Tetraodontiform fishes 132
        4.5.3.2 Poeciliid fishes 138

4.6 Allometric Growth and Shape-Change 142
    4.6.1 Shape-change of two atherinid fishes 144
        4.6.1.1 Two versions of the data base 144
        4.6.1.2 Shape-difference between species 146
        4.6.1.3 Growth of *Atherinella pachylepis* 148
        4.6.1.4 Growth of *Atherinella* sp. 151
        4.6.1.5 Explanations of shape-change 152
        4.6.1.6 Conclusions 152
    4.6.2 Ecophenotypic differentiation in a cottid 153
        4.6.2.1 Shape differences between ecophenotypes 153
        4.6.2.2 Shape differences between size-standardized forms 156
        4.6.2.3 Shape difference in relation to growth 157
    4.6.3 Shape differences in closely related species of *Cottus* 159
        4.6.3.1 Discrimination between species 159

4.7 The Components of Shape-Change 161
    4.7.1 Displacement and complexity 162
        4.7.1.1 Complexity as the scatter of a distribution 166
    4.7.2 Catostomid skulls 168
    4.7.3 Reading a complexity plot 169
    4.7.4 Phylogenetic applications 175
    4.7.5 Shape comparison of ontogenies 179

5 Geometric Morphometry and Evolutionary Biology 187

5.1 Nonstandard Homology Functions 188
   5.1.1 Duplication of homologues 188
   5.1.2 A hybrid morphometrics for nonsmoothness 189

5.2 Allometric Coherence and Reassociation 192
   5.2.1 Allometric coherence 193
      5.2.1.1 Growth gradients 193
      5.2.1.2 Morphological integration 194
      5.2.1.3 Unique variance 196
   5.2.2 Allometric reassociation 196
      5.2.2.1 Path analysis of reassociation 196
      5.2.2.2 Biorthogonal analysis of reassociation 197

5.3 Developmental Programs and Evolutionary Inference 198
   5.3.1 Survey of conventional terminology 199
   5.3.2 Format for quantitative investigation 200
   5.3.3 Without a time scale 202

5.4 The Investigation of Morphological Spaces 206
   5.4.1 Divergent evolution and taxonomic dissimilarity ... 206
   5.4.2 Examination of unoccupied morphological space 207

5.5 Recapitulation and Prospect 211

Appendices 215

A.1 Some Useful Coordinate Formulas 215

A.2 Biorthogonal Analysis of Triangles by Hand 218

A.3 The Variety of Coordinate Systems 225
   A.3.1 Other coordinates based in distances to lines 225
   A.3.2 Polar coordinates 226
   A.3.3 Mixed systems 226

A.4 Multivariate Analyses for a Triangle of Landmarks 230

A.5 Notes on Computations 238
   A.5.1 Three program listings 239

A.5.1.1 Principal directions and dilatations 239
A.5.1.2 A SAS procedure for shearing 243
A.5.1.3 Factor analysis using Wright's algorithm 245

Glossary 251
Literature Cited 263
Index 271

Preface

The origins of this book and the formation of the group of authors (the Morphometrics Study Group) lie in a course offered by Bookstein at the University of Michigan in the winter of 1980. After the end of that semester, the class continued meeting throughout the spring and summer, resulting in the first MSG contribution (Humphries et al., 1981). The following spring the MSG, now including Strauss, organized and participated in a symposium with others interested in morphometrics (American Society of Ichthyologists and Herpetologists, Corvallis, Oregon, 1981). Afterwards we decided to collect and publish our general approach to morphometrics and the new models and refinements of older methods that we were employing in our research.

The basic concepts of this book are discussed at some length in the introductory chapters; we strongly urge the reader not to attempt the later chapters without some familiarity with our basic tenets.

We envision this text as suitable for readers who have a working knowledge of biological statistics, including the concepts of regression, correlation, covariance, and principal components. For those whose last geometry class was in the distant past, we recommend *Geometry and the Imagination* (Hilbert and Cohn-Vossen, 1952) as a stimulating (re-)introduction to basic geometry.

Realizing that much of the material here will be difficult for the beginning morphometrician, we have provided an extensive glossary of technical terms, and hope that the reader will refer to it frequently. The first occurrence of each of its entries is noted in the text by boldface type. For example, the word **distance** will always be used to signify a physical length, measured perhaps in centimeters. General terms or phrases, such as principal component, are included in the glossary but are not in boldface in the text. Footnotes and appendices have been kept to a minimum; their contents are inessential for understanding techniques, results, or interpretations.

Some of the computer programs for techniques described here are presently dependent on features of the University of Michigan computing environment. At this writing, the programs are being assembled (by R.E.S. and F.L.B.) into an exportable statistics/morphometrics package. The authors welcome collaboration and encourage visitors interested in using the specialized geometric morphometry programs. See Appendices A.1, A.2, and A.5 for algorithms or instructions for manual computation. Appendix A.4 is a discussion of statistical testing of morphometric data.

All the geometrical analyses presented in this work are two-dimensional. This represents no theoretical constraints of our techniques, but instead reflects the programming effort necessary to add another dimension to the analyses.

Preface

We would like to express our gratitude to William Atchley, Ted Ladewski, Les Marcus, Stuart Poss, F. James Rohlf, and especially Sewall Wright for stimulating discussions concerning many of the topics covered in this book. We are grateful to Catherine Badgley, Robert Cashner, James Cheverud, William Fink, John Hendrickson, Robert Johnson, Les Marcus, James Mosimann, Melvin Moss, James S. Rogers, F. James Rohlf, and Donald Straney, who served as reviewers for most of the text. Our indebtedness to these colleagues does not imply their agreement with views expressed here. Responsibility for all errors of logic, omission, and presentation remains our own.

The authors are listed alphabetically; no attempt should be made to attribute particular sections to individuals, as every section received considerable input from all authors.

The production of this manuscript and most of the examples of data analysis were underwritten by NIDR grant DE-05410 to F. L. Bookstein. Computations were also supported by NSF grant DEB-8011562 to G. R. Smith and R. E. Strauss, by a University of Michigan Rackham Dissertation grant to J. M. Humphries, and by Edwin C. Hinsdale grants to both J. M. Humphries and B. Chernoff. The biorthogonal grid program (Sec. 4.5) and the symmetric axis program (Sec. 4.1) were written under NIDR grant DE-03610 to R. E. Moyers and NSF grant SOC-77-21102 to F. L. Bookstein. Current programming is supported by NSF grant BSR-8307719 to R. E. Strauss and F. L. Bookstein.

Ann Arbor, Michigan
August 1984

List of Tables

3.2.1 Strain statistics for the five trusses of Figure 3.2.4 59
4.1.1 Medial axis angles of bass opercles 80
4.2.1 Descriptive statistics for six measures of White Leghorn fowl 91
4.2.2 Principal components of the Leghorn correlation matrix 91
4.2.3 Loadings on a simple size factor, and residual correlations 91
4.2.4 Loadings and residuals on primary size after sequestering 94
4.2.5 Path coefficients for the Leghorn model with four factors 94
4.2.6 Analysis of size factors, Figure 4.2.10 96
4.2.7 Secondary factor, four pelvic distances, Figure 4.2.10 98
4.2.8 Analysis of size factors, Figure 4.2.11 99
4.3.1 Simulations of group discrimination with identical size distributions 112
4.3.2 Simulations of group discrimination with four intragroup size factors
      and differing size distributions 115
4.3.3 Coefficients for analyses of species of *Gila* 120
4.6.1 Principal component loadings of 22 conventional morphometric measures
      on *Cottus cognatus* 155
4.6.2 Principal component loadings of two different character sets for
      Cottus klamathensis and *C. pitensis* 160
4.7.1 Comparisons of shape differences among skulls of catostomid fishes 182
4.7.2 Ontogenetic comparisons of sucker skulls 185

List of Figures

1.2.1 The homology function 5
1.2.2 The geometric form 7
1.2.3 Landmarks must be characterized locally 8

2.1.1 *Diodon* and *Mola* 15
2.1.2 Deformations may be taken as linear over small regions 17
2.1.3 Biorthogonal directions bear the maximum and minimum diilatations 18
2.1.4 Demonstration of the existence of biorthogonal directions 20
2.1.5 The method of biorthogonal directions for two homologous triangles ... 21

2.2.1 Schematic of dual space for three variables 29
2.2.2 The plane of size vectors 30
2.2.3 The plane of shape vectors 30
2.2.4 The size vector $\tilde{S}$ 31
2.2.5 Correlations between shape and size variables 32
2.2.6 Unwanted correlations between shape and size 32
2.2.7 Three regression models for size-correction 33
2.2.8 Construction of shape variables as residuals 34
2.2.9 Instability of covariances among residuals 34

2.3.1 The ambiguity of homology-free analysis of outlines 39
2.3.2 Critique of the geometric analysis of shape variation 41

3.1.1 Four examples of more-or-less conventional sets of distance measures
      used in fish morphometrics 47
3.1.2 Loadings as descriptors of shape-change 48
3.1.3 Mapping Cartesian coordinates of points by triangulation 49
3.1.4 Examples of triangulation schemes for a set of seven landmarks 50
3.1.5 Cartesian coordinates as measured distances from points to each of a pair
      of perpendicular lines 51
3.1.6 Cartesian coordinates are not homologous distances 51

3.2.1 Possible patterns of distances which completely archive a configuration
      of ten coplanar landmarks 53
3.2.2 A truss for *Cottus* 55
3.2.3 The geometry of flattening one truss cell 56
3.2.4 Mapped landmark configurations for five specimens of the sculpin
      Cottus cognatus 58
3.2.5 Scatter plot of strain per individual against a composite measure of body
      size, for 43 specimens of *Cottus klamathensis* 61

List of Figures

3.2.6 Reconstructed truss networks for a single specimen of *Cottus cognatus* ..62
3.2.7 Scatter plots of the Cartesian coordinates of reconstructed truss networks for 10 specimens of *Cottus cognatus* 62
3.3.1 Points to be digitized 63
3.3.2 Varieties of boundary arc between landmarks 65
3.3.3 Schematic of the atherinid data base 66
3.3.4 Digitizing scheme for Figure 3.3.3 67

4.1.1 The medial axis 70
4.1.2 Triple points in a medial axis 71
4.1.3 A simple medial axis 71
4.1.4 Correspondence of anatomical regions and medial axes 73
4.1.5 Change in a medial axis with growth 74
4.1.6 Size series of opercular outlines with medial axes 75
4.1.7 Growth series of opercular outlines with medial axes 76
4.1.8 Superimposition of opercular outlines 77
4.1.9 Close approximation of triple points in medial axes 78
4.1.10 Angular measurement on a medial axis 79

4.2.1 Multiple regression as a path analysis 81
4.2.2 Direct and indirect effects 81
4.2.3 Path diagram for conventional discrimination 83
4.2.4 Corrected path analysis for discrimination 83
4.2.5 The two paths for a discriminant coefficient 83
4.2.6 Path diagram for size-free discrimination 84
4.2.7 Path diagram for a factor on two variables 87
4.2.8 Path diagram for a perfect factor loading 90
4.2.9 Path diagram for Wright's Leghorn data 95
4.2.10 Eleven distance measures for *Cottus* 100
4.2.11 Distances to endpoints of *Atherinella* 100

4.3.1 The geometric effect of shearing a scatter 104
4.3.2 Path diagram for a set of five simulations 111
4.3.3 Scatters from simulations of group discrimination with identical size distributions 113
4.3.4 Scatters from Simulation IV 116
4.3.5 Scatters from Simulation V 117
4.3.6 Shapes of young and adult members of three species of *Gila* 118
4.3.7 Principal component analyses of three species of *Gila* and their hybrids 119
4.3.8 Identification of unknowns 122

4.4.1 Averaged forms for two species of Cottus at three different standard body sizes 125
4.4.2 Dependency of strain on the size at which a truss is standardized ...126

List of Figures

4.5.1 Cartesian deformation of *Diodon* into *Mola* 129
4.5.2 Cartesian deformation of primate skulls 129
4.5.3 Biorthogonal analysis of the transformation of a square into a
 quadrilateral 131
4.5.4 Construction of an outline for biorthogonal analysis of tetraodontiform
 fishes 133
4.5.5 Biorthogonal analysis of the transformation of *Diodon* to *Mola* 134
4.5.6 Biorthogonal analysis of the caudal regions of *Diodon* and *Mola* 136
4.5.7 Biorthogonal analysis of the transformation of *Mola* into *Diodon* 137
4.5.8 Probable phylogenetic relationships among selected genera of
 tetraodontid fishes 138
4.5.9 Biorthogonal analysis of the transformation of *Ranzania* to *Masturus* ..139
4.5.10 Biorthogonal analysis of the tranformation of *Ranzania* to *Mola*...... 139
4.5.11 Cartesian deformation of poeciliids................................ 140
4.5.12 Construction of an outline for biorthogonal analysis of poeciliid fishes 141
4.5.13 Biorthogonal analysis of two digitizations of the outline of *Gambusia
 affinis* 142
4.5.14 Biorthogonal analysis of poeciliid fishes 143
4.5.15 Size-standardized biorthogonal analysis for selected pairs of poeciliid
 fishes 144

4.6.1 Radiograph of *Atherinella pachylepis*................................ 145
4.6.2 Analytic boundary of *Atherinella pachylepis* 145
4.6.3 Analytic boundary of *Atherinella pachylepis* with head, eye and fins
 added 145
4.6.4 Distance measures for analysis of atherinids........................ 146
4.6.5 Scatter of scores on first two principal components for *Atherinella* 147
4.6.6 Scatter of scores on shape factor and first principal component for two
 species of *Atherinella* 148
4.6.7 Discriminating distance measures for *Atherinella pachylepis* and *A.* sp. .149
4.6.8 Biorthogonal grid distortion from *Atherinella pachylepis* to *A.* sp. 149
4.6.9 Biorthogonal grids for growth in *Atherinella* 150
4.6.10 General size factors for *Atherinella pachylepis* and *A.* sp. 151
4.6.11 Principal components analyses of samples of the sculpin *Cottus
 cognatus* collected from lakes and streams 154
4.6.12 Loadings of the sheared second component of the truss data
 (Figure 4.6.11c) depicted on the truss network 156
4.6.13 Biorthogonal analysis of averaged trusses for lake and stream sculpins .157
4.6.14 Multivariate within-group allometric coefficients, depicted on the
 truss networks 158
4.6.15 Principal component analyses of two character sets on the same 64 specimens
 of the sculpins *Cottus pitensis* and *C. klamathensis* 161
4.6.16 The differences in size-free shape between *Cottus klamathensis* and
 C. pitensis 161

List of Figures

4.7.1 Log dilatations as coordinates 164
4.7.2 Scale- and proportion-change as Cartesian coordinates 165
4.7.3 Squared Euclidean distances in the scale-proportion coordinate system .. 166
4.7.4 Scatter of dilatations for a nonlinear transformation 167
4.7.5 Decomposition of sums of squares in the scale-proportion plane 168
4.7.6 Cladogram of catostomid genera 169
4.7.7 Digitizing scheme for catostomid skulls 170
4.7.8 Catostomid skulls 171
4.7.9 Shape comparison of *Erimyzon* and *Hypentelium* skulls 173
4.7.10 Shape comparison of *Chasmistes* and *Deltistes* skulls 174
4.7.11 Shape comparison of small and large *Moxostoma* skulls 176
4.7.12 Comparison of forms of sucker skulls with an outgroup form 178
4.7.13 Comparison of ontogentic transformation among *Erimyzon* and possible sister groups 180
4.7.14 Comparison of amounts of ontogenetic and phyletic change 184
4.7.15 Comparison of the complexities of ontogenetic and phyletic change ... 185

5.1.1 Ambiguities of homology 190
5.1.2 Two modes for dealing with nonconforming landmarks 191
5.1.3 Limits of the current deformation algorithm 192

5.2.1 Path-analytic factor model of allometric coherence 194

5.3.1 Diagrams for the discussion of heterochrony 200
5.3.2 The curve of shape-change as a function of size-change 203
5.3.3 Rates of change of shape-change with respect to size-change 205

5.4.1 Example of the reconstruction of body forms at unoccupied points in morphological space 208
5.4.2 Principal components II and III of the morphospace of seven genera of Pacific sculpins and the outgroup genus *Scorpaena* 210
5.4.3 A cladogram of the eight genera of Figure 5.4.2 211
5.4.4 The topology of the cladogram of Figure 5.4.3, superimposed on the projected morphospace of Figure 5.4.2 212

A.2.1 Biorthogonal analysis of triangles by hand, Step 1 218
A.2.2 Step 2 219
A.2.3 Step 3 220
A.2.4 Step 4 221
A.2.5 Step 5 221
A.2.6 Step 6 222
A.2.7 Step 7 223
A.2.8 Step 8 224
A.2.9 Step 9 224
A.2.10 Step 10 225

List of Figures

A.3.1 Size allometry of a conventional ratio 226
A.3.2 Confocal conics 227
A.3.3 A bicircular quartic coordinate system 227
A.3.4 A parabolic coordinate system 228
A.3.5 Orthogonal pencils of circles 228
A.3.6 A parabolic circular coordinate system 229
A.3.7 The "polar form," a coordinate system for lines 229
A.4.1 The superposition of two triangles on a common basaseline 230
A.4.2 Shape coordinates 231
A.4.3 A scatter in the plane of shape coordinates 232
A.4.4 The scatter of pairs of triangles 233
A.4.5 Invariance of the shape coordinate scatter 234
A.4.6 Test of a difference in mean shape 235
A.4.7 Reporting a mean shape difference 236
A.4.8 Size allometry in the shape coordinate plane 237

1

Introduction

1.1 Morphometrics in the Context of Evolutionary Biology

Evolutionary and systematic biologists are concerned with the discovery and explanation of differences among organisms. Their activities—discrimination of taxa, description of ontogenetic or evolutionary change, and testing of evolutionary hypotheses—frequently require the analysis of morphological patterns. Current methods for comparing biological forms range from classic verbal and pictorial representation to lists of measured distances between identifiable points on an organism. A third mode, assignment of coded values to qualitative character states, is a controversial hybrid. Where forms are complex, or identifiable points (e.g., landmarks) are not well defined, the classical approaches have the advantages of subtlety and flexibility unconstrained by the limitations of the more formal methods. Systematics based on coded character states enjoys some of the flexibility of the classical methods (while sharing their lack of rigor) and is also compatible with numeric methods for finding minimum-length trees.

In this book we suggest ways to compare forms for the discrimination of groups and the description of change. We avoid certain conceptual controversies in systematics and evolution—cladistics vs. phenetics, punctualism vs. gradualism, etc.—so as to address a problem common to all: the collection and manipulation of appropriate data on form differences. The more informative the data base and its analyses, the more effective will be the explanation that is achieved.

The techniques offered here are quantitative, after the tradition of numerical taxonomy, but our spirit is more classical in that we initiate analyses with particular comparisons in mind and extract as much information as the data contain with respect to those comparisons. However, we do not exhaust the data; different comparative contexts will lead to extraction of different subsets of the data. When possible, discriminatory and descriptive characteristics should be derived as products of analysis without the prior selection of favored characters. The even distribution of measures over a form, long an objective of numerical taxonomy, is approached here in a scheme of measures among landmarks in a pattern of quadrilaterals, the **box truss** (Sec. 3.2).

In contrast to most methods current in morphometrics, our evaluation of the effect of size on shape differences is achieved not by attempting to eliminate size, but by treating both size and shape as unmeasured **factors** implicit in the covariance among measurements: allometrically coupled aspects of growth and evolution. As a solution to the problem of discrimination among groups that differ in sizes of individuals, the **shear** method is offered. Here, again, we suggest choosing measured variables in a pattern aimed at broad coverage for shape analysis, while utilizing meristic and other non-shape characters separately.

Perhaps the most distinctive contrast between other approaches and the ideas presented here is our contention that shape differences can be measured more effectively than by subtraction of separate shape measures. The measurement of shape change can exploit the fundamentally different concept of **deformation**, as foreseen by D'Arcy Thompson. The method of **biorthogonal grids** (Sec. 4.5), an extension of Thompson's ideas, allows the quantitative comparison of forms by revealing the principal directions of form change. We use it to study ontogeny, size and shape, morphological complexity, and taxonomic divergence. It is also a helpful initial analytical tool in discovering characteristics of form variation for subsequent analysis.

The methods discussed below are being developed in light of rapidly growing technological possibilities for shape-analytic tools in a shape-rich field. Although most of the examples are two-dimensional, especially those that are image-mediated, the methods are intended to be generalized to three dimensions.

1.2 Selection of Models and Methods

In the course of research, the quantitative biologist must decide upon an endless series of algebraic or geometric details of procedure. The collective impact of these details upon the ultimate merit of the study can be surprisingly great. Morphometrics, the study of biological questions using geometric information, offers choices in two separate arenas: of geometric data and of statistical method. There results an unusual richness of methodological variation, a great empirical flexibility. Geometric data may be recorded as collections of measured distances, coordinates of landmark points, landmarks augmented by information about the curving of form between them, or information about curving of form without information about landmarks. (These choices are described in Chapter 3.) There may be other variables available more or less removed from the geometry—age, weight, sex, color, habitat, epoch, or any of a myriad of others. The report of a trend or contrast may, among other possibilities, take the form of a single distance or ratio measure, a pattern of coefficients on a principal component or factor, a single symmetric tensor (Sec. 2.1) or a field of them (Secs. 4.5–4.7), a graph down the middle of an outline (Sec. 4.1), or the octets of shape-difference scores described in Section 4.7.

We do not offer formal recommendations for the selection of a universal strategy from this spectrum. Rather, we describe in Section 1.2.3 a pair of fundamental

1.1 Selection of Models and Methods

concepts, the deformation model and the factor model, which in tandem may supply defensible rationales for particular choices in particular studies. Here are some examples of the sort of thinking to which they lead.

1. The notion of "shape" is often made operational using ratios of measured distances. When data are recorded by landmark coordinates, or in such a way that landmark coordinates can be recovered, the variety of possible ratios is immense. The ratios of distances at 90° then play a particularly important role (Sec. 2.1). When such data are not available, the set of analytic summaries is more tightly bound to the distances actually measured. In this case additional criteria about the interplay of shape variables with size variables (Sec. 2.2) come into play, criteria which are less crucial when the geometric data base is richer.

2. Information about curving form may or may not be accompanied by information about biological homology. The forms of quantitative summary corresponding to these two situations are quite different (Sec. 2.3). In general, if homology information is present, its relevance to a quantitative summary may be checked; if that information is absent, its effect upon the summary may not be presumed.

3. When a method involves specification of a particular registration, orientation or scaling of a geometric image, or any other special coordinate system or superposition, the specification should be recognized as an arbitrary choice and its impact upon subsequent statistical analyses explicitly investigated. (See Chapter 3.) Alternate forms of analysis might not require such a standardization, or might be used to determine whether the choices made lose information in some class of interesting applications.

With flexibility of method in mind, we will speculate about the sort of data to which a particular method might be suited, the phenomena which it can describe without loss of information, and the simplest checks for the existence of such lost information in particular applications. When in these pages we disapprove of particular morphometric studies, it is usually because the (inevitable) arbitrariness of the analytic steps pursued was not properly recognized or alternatives were not explored. Although no morphometric method can be wrong in all contexts, neither is any method universally applicable. Our primary theme is the matching of morphometric machinery, both geometrical and statistical, to the biological context: the nature of the information available about particular forms and the nature of the comparisons to which the study is, or could be, addressed.

1.2.1 Homology and homologous points

D'Arcy Thompson (1961: 274–275) introduces the study of **form-comparisons** in this way:

> ...The morphologist, when comparing one organism with another, describes the differences between them point by point, and 'character' by 'character.' If he is from time to time constrained to admit the existence of 'correlation' between characters ..., yet all the while he recognizes this factor of correlation somewhat vaguely, as a phenomenon due to causes which, except in rare instances, he can hardly hope to trace; and he falls readily into the habit of thinking and talking of evolution as though it had proceeded on the lines of his own descriptions, point by point, and character by character.
>
> ...But when the morphologist compares one animal with another, point by point or character by character, these are too often the mere outcome of artificial dissection and analysis. Rather is the living body one integral and indivisible whole, in which we cannot find, when we come to look for it, any strict dividing line even between the head and the body, the muscle and the tendon, the sinew and the bone. Characters which we have differentiated insist on integrating themselves again; and aspects of the organism are seen to be conjoined which only our mental analysis had put asunder. The coordinate diagram throws into relief the integral solidarity of the organism, and enables us to see how simple a certain kind of correlation is which had been apt to seem a subtle and a complex thing.
>
> But if, on the other hand, diverse and dissimilar fishes can be referred as a whole to identical functions of very different co-ordinate systems, this fact will of itself constitute a proof that variation has proceeded on definite and orderly lines, that a comprehensive 'law of growth' has pervaded the whole structure in its integrity, and that some more or less simple and recognisable system of forces has been in control.

In this manner Thompson explicitly identifies the object under discussion, namely the relationship between biological forms, with a **latent variable**: an abstraction designed to explain diverse comparative findings distributed over the organism. His identification of this latent variable with a "system of forces" reflects the biomechanical understanding typical of his era, now obsolete; what remains, the idea of a "simple and recognisable" geometrical pattern of explanation (Fig. 1.2.1), has fascinated mathematical biologists and morphologists from Thompson's day to the present.

The endurance of Thompson's insight owes to its felicitous blending of two previously unrelated descriptive traditions. The latent variable of which he speaks had hitherto been studied in biology and in mathematics separately, where it went by two different names. The biologist knew it as **homology**, the rules by which parts of different organisms were understood to correspond, whereas the mathematician knew it as the pointwise **deformation**, or "Cartesian transformation," acting to distort a picture or other specifically geometric representation of form.

In saying that deformation is a mathematical model for homology, as Thompson did in the quotation above, we mean that the quantitative statistical analysis of

1.2 Selection of Models and Methods

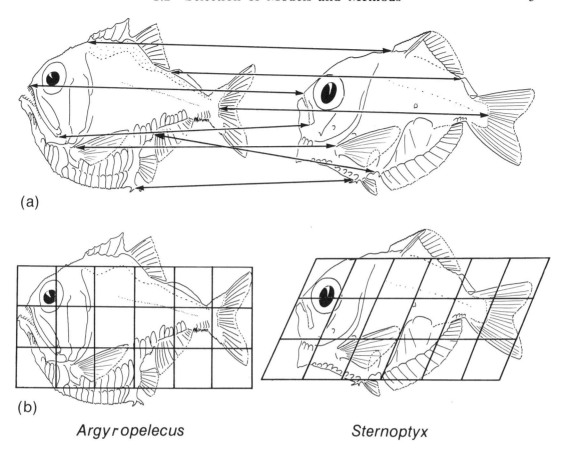

Argyropelecus   *Sternoptyx*

Figure 1.2.1 The homology function. The homology function is a mathematical model of homology as a smooth deformation. One way of displaying the correspondence of homologous points (a) is by a Cartesian grid transformation (b). Example after Thompson (1961).

deformation, according to the methods set forth here, provides geometrical summaries of differences that will eventually be "explained" biologically. Whenever we find systematic aspects of deformation that correspond to trends or contrasts of biological interest, we assume that the biological interpretation of these findings will crucially involve phylogenetic or ontogenetic aspects of homology (Chapter 5).

Biological homology refers to the rules by which definable structures or "parts" correspond: we speak of the homology of the reptilian quadrate and the mammalian incus. It is not the purpose of this manuscript to discuss the rules—spatial, ontogenetic or phylogenetic—by which these correspondences are founded. We will, however, extend the traditional use of homology to include the correspondence of points and boundaries as well as parts. For example, we will consider the origin of the spinous dorsal fin in two centrarchid fishes to be a homologous point. Likewise, the segment of the boundary between the origin of the spinous dorsal and the origin of the soft dorsal will be considered homologous in those two forms. (For the computation of homology in this sense, see Section 3.3.2.) The definition and verification of such homologies are identical to those of their more traditional counterparts. We will refer to points whose comparisons are consistent with the rules of homology and that have reliable anatomical definitions as **landmarks**.

Because we are interested in comparisons of forms rather than their separate definitions, we introduce here the concept of the **homology function** or **homology map** (Woodger, 1945; Jardine, 1967, 1969; Bookstein, 1978a:120–123), a geometric tool for describing the correspondence between two or more forms. Specifically, a homology function is a mathematical **model** that starts with the landmarks and smooth boundary curves of one form and describes how to map these onto the homologous curves and points of a second form. This mapping is modeled as **smooth**: points nearby in one form correspond to points relatively nearby in all other forms, and short segments nearly straight in one form do not manifest large kinks in other forms.

The homology function associates different kinds of points—landmark and non-landmark—among geometric forms. (By **geometric form** (Fig. 1.2.2) we mean the landmark points and curving arcs between them that constitute a sketch of the organ or organism under discussion.) We will refer to **computed-homology** as the relationship among points (one on each geometric form) that are interpolated between appropriate landmarks. Included in this definition will be all interior points not located by reason of their anatomical definition, as well as specific points along the boundary—for example, the point halfway between the origins of the spinous and soft dorsal fins. In accordance with this distinction we will speak of two types of correspondence: **homologues**, which are defined by the conventional rules of biological homology and which drive the mapping function, and **computed-homologues**, which are defined by the mapping function.

The biological aspect of this homology function, as it associates points or curves, varies from example to example. When we study a fish scale or a mollusk shell with growth rings, for instance, the approximate history of the organ and its homologous points and curves may be explicit in its image at any moment. Sometimes, instead, homologues are analogous by position: the sutures separating a particular pair of bones should be homologous as curves from form to form of a skull series. Intersections of sutures often serve to characterize homologous points.

Computation of a homology function from data representing two forms begins with a sample of the correspondence: **landmark** sets, which must be located reliably in all the forms of a series. The computed-homologues for all the other points—the non-landmarks—of a form are interpolated among the locations of the homologous landmarks.

Landmarks must satisfy several constraints imposed by the deformation model. Although landmarks may later be assigned coordinate values, a good landmark needs to have an operational definition in terms of the anatomy and/or ontogenetic history in its vicinity. It is not sufficient to choose as landmarks points having extreme values relative to a particular coordinate system. For example, the designation of highest or

1.2 Selection of Models and Methods 7

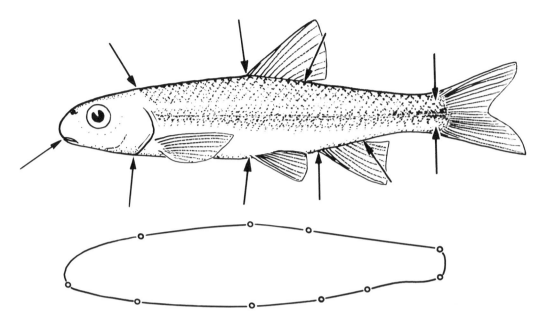

Figure 1.2.2 The geometric form. The geometric form represents a biological specimen by the set of landmarks and curving arcs between them. Landmarks are indicated by arrows on the drawing and by dots on the geometric form. In this example we have excluded the fins.

lowest points of the form in Figure 1.2.3 as supposed homologues is arbitrarily dependent on the orientation of the form relative to the coordinate axis, here the line from premaxilla to center of hypural plate. When a coordinate system is defined by reference to landmarks, it is inappropriate to subsequently define landmarks in terms of their coordinates, because inconsistency in positioning the forms can cause dramatic shifts in the location of these supposedly homologous points. Other ways of defining one landmark in terms of another (i.e., endpoints of minimum or maximum diameters) may also lead to violations of homology. For example, points constructed as nearest to or farthest from other landmarks, and geometric combinations of landmarks at a distance, are not proper landmarks. These criteria proscribe intersections of lines or "axes," points at extreme distance from landmarks defined earlier, and so forth. "The base of the dorsal fin" will be definitive on many fishes; "the highest point of the back" will not. Points on a projected drawing or radiograph representing the abutment of three structures in space are often helpful selections; points where two curves cross in projection but not in space are not as useful.

The homology function, anchored at the landmarks, specifies a computed correspondence along the outline arcs and within the interiors of the forms. This requires that the landmarks be selected not individually but as a set, over all forms, to

8   1 INTRODUCTION

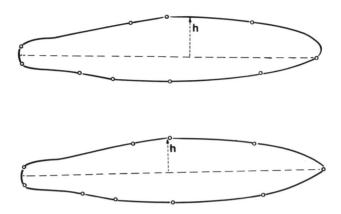

Figure 1.2.3 Landmarks must be characterized locally. As we change the form of the fish, the spatial relationships of "the highest point of the back" alter excessively. Here "height" is distance above a segment from the premaxilla to the center of the hypural plate.

ensure that the correspondence as computed is sensible (see Sec. 5.1). In practice, certain landmarks with a clear anatomic definition may have to be excluded from a morphometric analysis by this criterion. For instance, landmarks very close together in some forms should not be far apart in others; and landmarks found within a region defined by other landmarks in one form should not be outside of that region in another form. While such landmarks may be appropriate in multivariate analyses of morphometric distances, they are unsuitable for morphometric modeling of shape-change as a deformation.

Landmarks not only anchor the homology maps but also guide the construction of data sets for samples of geometric forms. For this purpose, landmarks may be assigned **coordinates** in any convenient coordinate system (Cartesian, polar) registered and oriented howsoever. In place of coordinates, we may record any other system of quantities that permits the complete reconstruction of the landmark configuration, such as the box truss of interpoint distances (Sec. 3.2). Note that the **archive** of coordinate data contains no derived variables: no angles, ratios, diameters, functions of azimuth, or other complex constructs.

Arcs of the outline between landmarks may be recorded by sequences of points (i.e., coordinate pairs) along them or instead may be abstracted into curves represented by formula—arcs of circles, ellipses, and so forth (see Sec. 3.3). Whether such a representation, or any particular selection of landmarks and connections, is satisfactory depends entirely on the empirical context of the research and the comparisons that are to be investigated. For certain comparisons information on boundary arcs may be discarded, whereupon the outline is represented by the straight lines connecting the landmarks around the form.

1.2 Selection of Models and Methods

1.2.2 Computing variables from comparisons

The coordinate data from the morphometric archive—coordinates or parameters of arcs, form by form—are not intended to be used as statistical "variables." Multivariate analyses of these coordinates or their ratios are not appropriate for description of trends or contrasts of size and shape (Sec. 3.1.3). Even multivariate analyses of interlandmark distances may be less than optimal when the distances fail to align with the principal directions of the form comparison. From a morphometric archive, interpreted as a collection of outline forms together with a biological homology function relating each to all the others, one might compute many arbitrary measures form by form. For instance one might select any two points, landmarks or not, upon any single form and measure the distance between their homologues on every form; or select any three points and compute for every form the angle subtended by two of them at the third; and so forth. Of the indefinitely many meaningfully comparable variables that could result, all analyses of finite subsets chosen a priori will discard some of the information of the archive in unknown amounts, although some subsets are intrinsically less wasteful than others (Sec. 3.2).

Usually, a character set of satisfactory structure can be extracted only in a specific empirical context: the description or corroboration of contrasts between groups, trends within groups, contrasts controlling for trends, or diagnoses of the deviations of individuals from groups. Each of these tasks can be accomplished using the basic coordinate archive alone (forms plus homology maps) without restricting oneself to any predetermined set of variables or characters. In each context (e.g., contrast or trend) the initial findings indicate particular variables that ought to be extracted, variables likely to be optimal for the particular explanatory task at hand (see Sec. 2.1).

The conventional multivariate approach proceeds from a list of characters to optimal linear combinations of those variables for "explaining the most variance," "yielding the best separation," and the like. As an alternative, we propose that the form comparison of interest be objectively described using only the data base of location and homology. The resulting description will generate variables for the specific explanatory purpose, characters bearing information about size change and shape change as explicitly as possible, given the data.

The assembly of one fixed set of characters to serve in the investigation of a wide variety of hypotheses is likely to be misleading; in contrast, reserving until last the definition of variables proves an effective use of the information in the data base. For example, at any point inside an outline of the form, the biorthogonal method produces the proportion that shows the greatest change from that form to another. The description of such a comparison should be based in these proportions, whose construction cannot be inferred from any characteristics of the forms separately. Likewise, for the other kind of morphometric archive, that of distances, **factors** (see the next section) serve this role; they are "unmeasured" variables estimated near the end of an analysis. We define **general size**, for example, as a factor best accounting for empirical covariances among alternate "size measures." This factor leads, via Sewall

Wright's path-analytic methods, to a whole array of derived indices of size, each embodying potentially interesting contrasts within a single population.

1.2.3 Two basic models for morphometrics

The morphometric tools in this book serve for a kind of geometric illumination or corroboration of contrasts and trends of interest: hypotheses relating size-change, shape-change, and group membership. Examination of these hypotheses is crucially bound up with **morphometric models**. By these we mean geometrical/statistical machinery by which summary trends or contrasts, in combination with statistical noise, reconstruct (that is, "explain") the original data. Whenever we refer to "models," we mean morphometric models, unless the term is prefixed otherwise. These models always include an error term. Other techniques, such as principal components analysis, are transformations of data without error terms and will not be referred to as models.

There are two fundamental models underlying the techniques and examples of this book: form-change as deformation, and size and shape as factors of distance variables.

1.2.3.1 Form-change as deformation.—The Cartesian grid of D'Arcy Thompson (1961) is an explicit visualization of the homology map. Thompson did not complete the mathematical development of this object into a formal **tensor field**—that was done by Richards and Kavanagh (1943) and Bookstein (1978a). The tensor field assigns a ratio of change, in units of mm/mm, to every finite distance measurable upon a form and its homologue (Sec. 2.1).

To measure change of form, it is insufficient to measure the forms separately and subtract homologous measures. Rather, **form-change** is an object of measurement in its own right, independent of the measurable features of the related forms separately. The biorthogonal method starts by computing the homology map, then characterizes the map by the distances that change most and least quickly. For instance, "sexual dimorphism" will be the description of the transformation between a "typical" male and a "typical" female. After inspection of this description we may derive a collection of measures of separate forms, such as the optimal proportions already outlined above, in terms of which the difference between the forms is summarized more effectively than by any other set of variables.

A second purpose is served by these maps: they may be manipulated, compounded, and used for simulation. For instance, we can average form-changes, or apply form-change computed between two forms to extrapolate a third form, or to deform a fourth into a hypothetical fifth (see Secs. 4.4, 4.6.2).

This general deformation model is capable of mapping any configuration of landmarks onto any other. Most other techniques for the geometric study of form change restrict the relative displacements of landmarks, and thus are only special

1.2 Selection of Models and Methods

cases. For instance, one model constrains the change of the boundary to be toward or away from a presumptive "center" (see Sec. 2.3). Such a model can fail to explain much of the data (Moss et al., 1983; Bookstein, 1981c, 1983). Another specialized model (Siegel and Benson, 1982; Benson, Chapman, and Siegel, 1982) postulates a conserved subset of landmarks, a region of the form in which the transformation is nearly a pure change of scale. This model cannot recognize allometries (such as growth gradients) that vary smoothly by either position or direction. Yet another technique (Tobler, 1977, and references therein) employs a deformation that is homogeneous throughout the form. This transformation, which can be fitted by least-squares techniques, removes spatial gradients of allometry but admits directional differences.

In the specialized models, particular landmarks are often not to be found at the locations assigned to them by the fitted "deformation." That is, these models systematically fail to fit the observed correspondence of landmarks. The set of these discrepancies is itself a deformation, and could be analyzed as such by the methods of this book. For instance, it is inappropriate to invoke for any metric purpose the separate vectors that connect the two positions (modeled and observed) of the individual landmarks. Like any other vector method (Bookstein, 1978a, 1982a), this depiction requires that two configurations originally measured in separate coordinate systems be viewed in a common coordinate frame; the necessary superposition only rarely has any biological or biometrical reality insofar as it was arbitrarily selected from a vast number of arbitrary possibilities. In our view, none of the special versions of deformation analysis are worth computing separately. They are summaries only of certain special cases of deformation.

1.2.3.2 Size and shape as factors.—The deformation model assigns a ratio of change to every homologous distance variable. We will often consider multiple comparisons, contrasts, or trends, that is to say, multiple deformations (sex effects, age effects, etc.) jointly acting upon a morphotype or *Bauplan* within a single population. Each form is considered to be a specific deformation of the "typical" or mean, a deformation combined from the systematic effects that are the main comparisons: a certain amount of size-effect, a single dose of sex-effect or shape-effect, etc. These main effects each influence all possible distances, whether or not actually measured. Our models for the measured distances must therefore have the ability to account for the unmeasured distances as well. The only conventional multivariate models that serve this function are those of **factor analysis**. Therefore we model the effect of multiple deformations on populations by identifying the deformations (size effects, shape effects) with factors (size factors, shape factors) accounting for distances as in the classic allometric models (Wright, Huxley, Jolicoeur). In these models all distances, whether observed or not, are estimated as predicted values from regressions on linear combinations of **factor scores**; the coefficients of the combinations are called **loadings** (see Sec. 4.2.3). We will search among these loadings for patterns expressing the geometric organization of the measured distances, an organization that we interpret in biological terms.

2

Foundations

The morphometric practices we recommend generally involve landmarks, points reliably and homologously located in each form of a data set. The landmark is the empirical tie between the two kinds of information, geometric and biological, that ought to be conjoined in morphometric analyses.

For comparing configurations of landmarks, we will use quantities embodying the explanatory biological purpose of empirical studies. From form-*comparisons*, which are biological observations, come rules for designing and computing morphological variables. This interplay between deformations and variables, the fundamental analytic strategy in morphometrics, is the subject of the present chapter.

Section 2.1 explains how to describe the simplest changes of form by determining three particularly helpful variables: two distances that bear the greatest and least ratios of length between the forms, and one shape variable—the ratio of the same two distances taken within either form separately. These distances are perpendicular to one another in each of the forms being compared. Because this progression from form-change to variables is so important, we present two separate explanations of it, one more algebraic (Sec. 2.1.2) and one more geometric (Sec. 2.1.3). The practical aspects of producing these variables, by hand or by computer, are treated in Appendices A.2, A.4, and A.5.1.1.

Inasmuch as certain size and shape variables are generated intrinsically in the course of form-comparisons, it is appropriate to reflect critically upon the way in which such variables are ordinarily selected. Section 2.2 reviews the sets of size and shape variables arising in conventional morphometric studies. We consider two practices in particular: the transformation of distance measures to the logarithm (Sec. 2.2.1) and the adjustment of observed shape differences for size differences, variable by variable (Sec. 2.2.2). The former tactic, but not the latter, is consistent with the origin of size and shape variables in the comparison of forms.

Use of size variables that are distances between landmarks invokes the two kinds of morphometric information, geometric and biological, with equal force. Section 2.3 considers another collection of size variables, the "radius" or "width" function defined without much reference to biological homology. We shall indicate diagramatically that the information such analyses omit is crucial to the systematic study of biological shape change.

2 FOUNDATIONS

2.1 Rates of Change at a Point

2.1.1 General propositions

1. The measurement of biological **form-change** is quite different from the measurement of shape. Changes need not be quantified by measuring shapes separately and then subtracting scores on corresponding measures. Rather, form-change is a geometric object in its own right, the deformation of one form into another that accords with biological homology.

2. Of the information relevant to the analysis of shape change, that which is biological is manifested in the **homology map**, a function that assigns a correspondence between the points of any pair of forms. We generally sample this function at **landmarks**, points reliably located by anatomical criteria.

3. The choice of the coordinate system to be superimposed over any single form is not important. The computation of shape change is driven instead by the relation between two coordinate systems for two homologous configurations of points.

4. For sufficiently small regions, shape changes may be modeled as uniform. They are described by the rate of change of length, or **dilatation**, as a function of direction. Dilatation is measured as a ratio, not a difference, of homologous lengths. It is dimensionless.

5. The biorthogonal method summarizes the dilatations at any point by the largest and smallest, called the **principal dilatations**. The directions along which they lie, the **principal axes** or **biorthogonal directions**, are at exactly 90° both before and after transformation.

2.1.2 Shape change as a relation of coordinate systems

The measurement of shape change may be thought of as the description of one picture in the coordinate system of another.

2.1.2.1 The Cartesian grid.—Consider the famous transformation between the forms of *Diodon* and *Mola* which D'Arcy Thompson uses to introduce his method. (These drawings are really projections onto the midsagittal planes of the fish. The component of deformation that this operation contributes will be ignored until Chapter 5.) From the pair of forms alone, drawn in Figure 2.1.1a, it is clear that they differ a great deal. Thompson suggested that we construe this pair of forms as a transformation of the entire picture (plane projection) making homologous points correspond—taking mouth to mouth, eye to eye, tail to tail. We could have drawn this mapping with any set of points in the picture plane of *Diodon*.

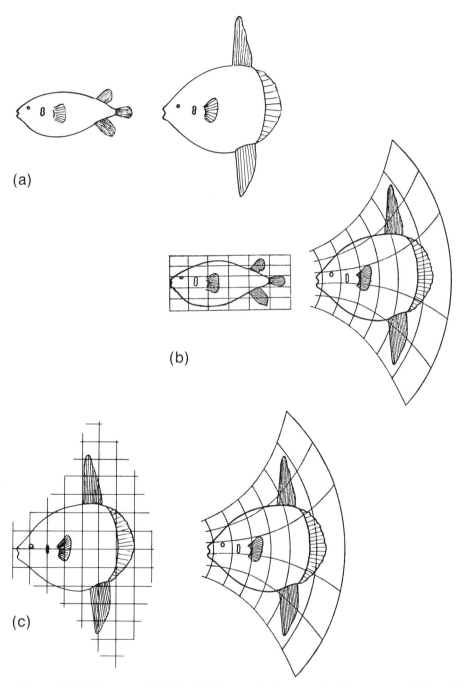

Figure 2.1.1 *Diodon* and *Mola*. (a) Geometric forms, after Thompson (1961), with a sampling of landmarks. (b) A Cartesian grid on *Diodon* and its transformation according to Thompson. Notice that the grid does not quite conform with the landmark homologies. (c) Alternate graphical representation of the data base: two coordinate systems upon the diagram of *Mola*.

To us this representation emphasizes numerical features of the form that are altered—lengths, heights, angles. This is a distraction. Thompson suggested that one separate the description of change from the description of the form, by superimposing an artificial structure over the drawing. In the course of deforming the fish, the homology map will deform the abstract structure as well. It is easier to describe the shape change by considering its effect on the abstraction, which is regular, than on the fish, which is not. For his abstraction Thompson chose to use an ordinary square ("Cartesian") grid aligned with the body axis of the left-hand fish. The grid is deformed (Fig. 2.1.1b) into a general curvilinear mesh that has no exact metric regularities. (The replacement of the extended Cartesian grid with one more appropriate to the task at hand, which is the description of shape change, will concern us in Section 4.5.)

Two different coordinate systems may be superimposed over the form on the right. One system is that drawn in Figure 2.1.1b, the deformation of the system that was Cartesian for *Diodon*; the other is the Cartesian system for *Mola* as the starting form, drawn as Figure 2.1.1c. The deformation of *Diodon* into *Mola* can be described as well by the relation between these two coordinate systems on *Mola*; all the information we need, both geometrical and biological, is still present.

The relation of the forms may be described most easily in small regions of the figure within which we can ignore the curving of the grid lines in *Mola*. Sufficiently small squares of a Cartesian grid will be transformed into parallelograms, as in Figure 2.1.2. The smaller the region, the better the approximation. This transformation takes any small straight line of tissue in *Diodon* into the homologous small straight line in *Mola*. These arbitrarily small straight-line segments are called **line-elements**. It is in terms of their correspondence that we shall describe the whole transformation.

2.1.2.2 Principal axes computed from coordinate mesh.—In the transformation of square into parallelogram both distances and angles are altered. We may begin our inquiry into the quantification of the transform by considering a specific query: Are there homologous pairs of line-elements that are at 90° in both forms?

What could we do if we had two such pairs of perpendicular line-elements, one pair in *Diodon*, one pair in *Mola*? We could learn a great deal about dilatations (the ratios measuring rate of change of length) in *every* direction in terms of the dilatations along just those two, as follows.

Let us assign a Cartesian center (0,0) in the middle of the square in the first form, and a local Cartesian coordinate system aligned with the axes we hope to have found, the **principal axes**, which start and finish at 90° (Fig. 2.1.3a). In this coordinate system, when we assign points coordinate pairs (x,y) relative to (0,0), their squared distance from (0,0) is $x^2 + y^2$.

We can assign coordinates on *Mola* that are set to (0,0) at the homologue of *Diodon*'s (0,0) and that are aligned with the two principal axes in *Mola*. Call these coordinates x' and y' (Fig. 2.1.3b). If d_1, d_2 are the dilatations along the principal axes, then we will have $x' = d_1 x$ and $y' = d_2 y$, by the uniform nature of the deformation in

2.1 Rates of Change at a Point

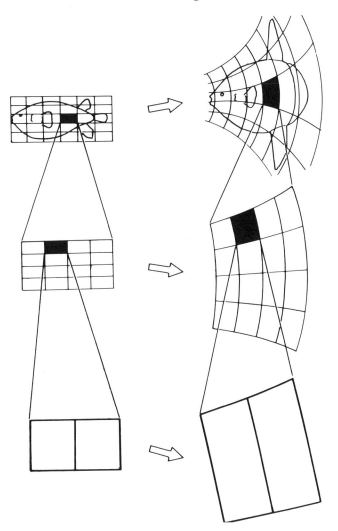

Figure 2.1.2 Deformations may be taken as linear over small regions. Successive enlargements of the transformation of *Diodon* to *Mola* show how small squares are transformed approximately into parallelograms.

this small region. The distance from the origin to (x',y') in *Mola*, then, will be $x'^2 + y'^2$, or $d_1^2 x^2 + d_2^2 y^2$. Let us assume that d_1 is greater than d_2; if not, we will simply reverse the subscripts.

Suppose we take the collection of points (x,y) all at the same distance from the origin in *Diodon*. These will be the points on a circle (Fig. 2.1.3c) of some convenient radius—call it 1 unit. The coordinates of these points satisfy the equation $x^2 + y^2 = 1$ or $y^2 = 1 - x^2$. The squared distance from $(0,0)$ to the homologue (x',y') in *Mola* becomes, with this substitution,

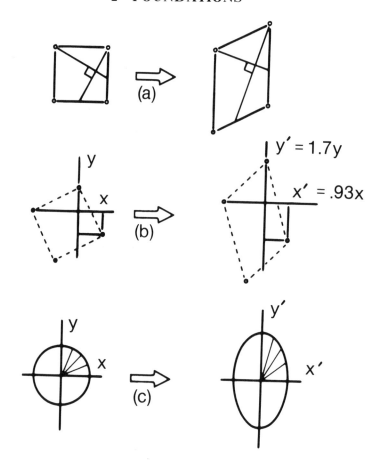

Figure 2.1.3 Biorthogonal directions bear the maximum and minimum dilatations. (a) Directions that start and finish at 90° for a transformation of the square into a parallelogram (cf. Figure 2.1.2 bottom). (b) Two Cartesian coordinate systems (x,y), (x',y') oriented with these principal directions, and then rotated to an arbitrary vertical. (c) A circle and its deformation. Dilatations may be represented by the lengths into which the deformation takes radii of a circle. This length is greatest along the principal axis of larger dilatation, and least along the principal axis of smaller dilatation.

$$d_1^2 x^2 + d_2^2(1 - x^2)$$

which is the same as

$$d_2^2 + x^2(d_1^2 - d_2^2).$$

Because in *Diodon* distances from $(0,0)$ to these points (x,y) were all the same, namely, 1, the distances from $(0,0)$ to their homologues (x',y') explicitly represent the factors by which distance has been expanded in the various directions. Because $d_1 > d_2$, by assumption, this *Mola*-distance will be greatest when x^2 is greatest. That happens for

2.1 Rates of Change at a Point

$x^2 = 1$, $x = \pm 1$, $y = 0$. These are just the points on the circle in the direction of the line-element we are temporarily considering to be horizontal (Fig. 2.1.3c). *Hence dilatations are greatest in this direction.*

Likewise, the distance from (0,0) to (x',y') is least, and so dilatations are least, for those points on $x^2 + y^2 = 1$ with $x^2 = 0$, the points along the other principal line-element, drawn as vertical in Figure 2.1.3c.

Thus the **biorthogonal** (bi-orthogonal) **directions**—the line-elements that start and finish at 90°—are the directions of greatest or least local rate of change of length out of all directions at a point. These rates are the **principal dilatations**, lying along the **principal directions** of the deformation for this small region. This dual property accounts for their remarkable descriptive efficiency.

We have explained the reason for desiring these line-elements; it would be very convenient if we were certain that we could find them. Our guarantee is supplied by an ancient theorem, one of the simpler proofs of which we will now present.

Let us consider the relation between angles in *Diodon* and in *Mola* by taking a pair of line-elements at a constant angle in *Diodon* and examining the angle that their homologues make in *Mola*. We take, in particular, the line-elements along our arbitrary Cartesian vertical and horizontal there. These lines are at 90° in *Diodon* and make an angle of some value or other in *Mola*, as shown in Figure 2.1.4. We may as well assume this angle is less than 90°, as drawn.

When we rotate this pair of elements in *Diodon*, keeping it always at 90°, its homologues in *Mola* will rotate, too, within its little homologous region. As they rotate, the angle between them will change as a smooth function of orientation in *Diodon*.

By the time we have rotated by 90° in *Diodon*, we have reverted to the same pair of line-elements with which we began, but now one of them has had its direction reversed. The angle which had been less than 90° in *Mola* has now been replaced by its supplement, which is of course greater than 90°.

Because this angle in *Mola* is a continuous function of the position of the right angle in *Diodon*, however, somewhere in-between it must have had exactly the value 90°.

This unexpectedly curt line of reasoning proves the existence of the biorthogonal directions. For another version of the same argument, see the Appendix on the Shape Nonmonotonicity Theorem in Bookstein (1980a). A slight extension of the algebraic argument above is sufficient to prove that these directions are unique: see Bookstein (1978a:103–104).

2.1.3 Shape change as a deformation of points considered in one coordinate system

In the preceding exposition we have argued as if the biological substance remained the same—i.e., the homology map was an identity—but the distance function was somehow changed; our task was to describe the directional variation of

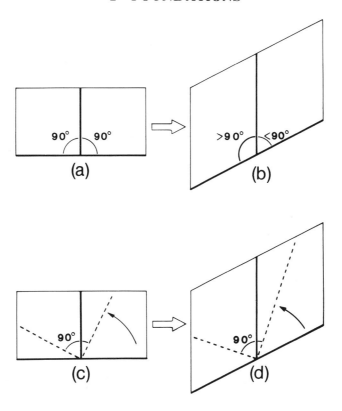

Figure 2.1.4 Demonstration of the existence of biorthogonal directions. One corner of the square (a) is deformed into an acute angle, another to an obtuse angle (b). As the pair of perpendicular lines in (c) rotates through 90°, the homologous lines in (d) rotate at different rates. The perpendicular pair in (c) must pass through one position for which its homologue spans an angle of 90° in (d).

those two distance-measures of the "same" line-element. One can arrive at the very same conclusion—the existence of biorthogonal grids at 90°, one direction bearing the largest dilatation, the other the smallest—by reversing the logical roles of the two coordinate systems. You will see that the use of Cartesian coordinates is *entirely dispensable* in the discussion of the principal axes and directions.

2.1.3.1 Line-elements in a triangle.—For the previous development, with its emphasis upon Cartesian coordinates, our basic unit of analysis was a bit of abstraction—a little square of coordinate mesh. For the alternate presentation our basic unit will instead be a homologous pair of triangles of landmarks, as in Figure 2.1.5a. In the absence of other information we may take the transformation sampled by these limited data to be uniform between each homologous pair of edges and throughout the interiors of the triangles. The homogeneity of this **affine transformation** is indicated clearly in the transformation grid Thompson-style (Fig. 2.1.5a).

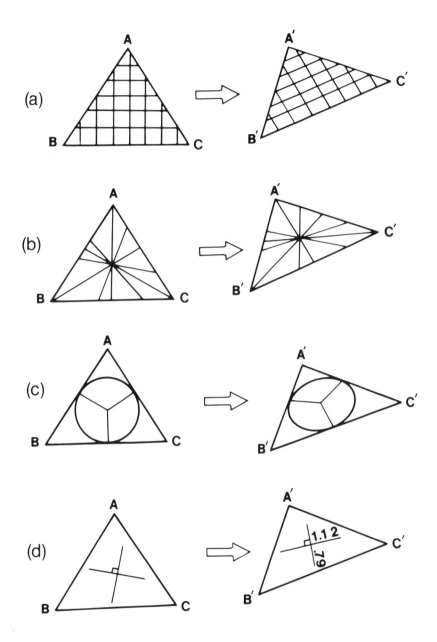

Figure 2.1.5 The method of biorthogonal directions for two homologous triangles. (a) The sets of landmarks suggest a uniform transformation of the interior. (b) The uniform transformation alters lengths in various directions. (c) Dilatations are proportional to radii of the ellipse into which a circle is deformed. (d) The principal directions are axes of this ellipse, and the principal dilatations are proportional to their lengths. The principal axes upon the circle, left, are along the diameters transformed into the principal axes of the ellipse.

But we may draw the transformation just as clearly in terms of the collection of lines in all directions (Fig. 2.1.5b).† The deformation we are observing, driven by the displacements of those landmarks at the corners, will deform these lines into others that divide the edges in the same fractions. That is, the deformation takes edges to edges, median lines (dividing the opposite sides in the ratio 50:50) to medians, and so on.

2.1.3.2 Principal axes computed from triangles of landmarks.—We are interested in the ratios of lengths of corresponding lines in the two triangles, the dilatations. We could compute them explicitly by taking quotients of corresponding lengths, direction by direction. However, it is more elegant to borrow the device of the previous demonstration. Again we may observe the dilatations directly as the lengths of the deformations of lines of constant length, that is, of radii of a circle. In this coordinate-free presentation we have the advantage of our geometric intuition. We can actually *draw* the circle (Fig. 2.1.5c) whose deformation we wish to observe, and the oval into which the uniform shear takes it. The dilatations of line-elements are proportional to the radii of this oval.

You may be convinced by an accurate drawing that this oval is remarkably like an ellipse. It is possible to prove by entirely non-algebraic (i.e., coordinate-free) means that it is exactly so. The image of the circle, being an ellipse, has two axes of symmetry, which lie at 90°. One is the largest diameter of the ellipse, one the smallest. The diameters of the circle that transform into them are likewise at 90°.

Recall that the lengths of the radii embody the dilatations as a function of direction. Therefore the principal axes of the ellipse into which a circle is taken *are* the principal directions of the deformation as they lie upon the right-hand form. The diameters that were mapped into them are determined by corresponding fractions of intersection along edges of the triangles. In Figure 2.1.5d we have drawn them without their ovals, showing that the remaining information about dilatations in intermediate directions may be readily reconstructed. The dilatations indicated on the figure were computed by division of lengths of homologous segments; all the other dilatations may be computed from these two alone, according to the formula

$$d_\theta^2 = d_1^2 \cos^2\theta + d_2^2 \sin^2\theta$$

where θ is the angle between the direction and the axis of larger principal dilatation.

We thus arrive at the same conclusion by two rather different lines of argument. Our unit of analysis may be the abstract cell of coordinate mesh or the biologically real triangle of landmarks; our line-elements may be bloodless little vectors inside parallelograms or real segments between points lying in fixed relation to those

---

† By an argument that we will not present here (see Bookstein, 1982c: part ii), dilatations for affine transformations are a function of direction only, so that we may restrict our attention to lines all through a convenient center.

2.2 The Derived Data of Size and Shape

landmarks. The conclusions are the same in either demonstration, because they follow rigorously from the identification of **biological homology** with **geometrical deformation**. At every point of a deformation representing biological homology there are two directions that are at 90° both before and after deformation; one is along the direction of greatest rate of change of length between homologues, and one of least rate, of all directions at that point. These directions, a description of the form-comparison, involve both forms for their computation, and cannot be determined by consideration of either form separately.

There is one exception to this elegant simplification. The ellipse at the right of Figure 2.1.5d may be a circle—it may not have a unique pair of axes. In other words, the dilatations d_1 and d_2 in the algebra of the previous section may be equal. In this case, dilatation is not a function of direction. We already know the deformations to which that description applies: they are the simple changes of scale. Such transformations may be represented locally by any pair of lines at 90°, because all right angles, indeed all angles, are preserved. These special transformations are called **isotropic**; points at which the transformation is isotropic are called **singularities** of the biorthogonal description. We shall encounter them from time to time as special cases of the more general deformation.

Appendix A.5.1.1 lists a Fortran subroutine that extracts the principal directions and dilatations relating two sets of three points in the Cartesian plane.

2.2 The Derived Data of Size and Shape

Throughout this book our data are those that conform to an underlying model of smooth deformation or transformation: (1) locations of homologous landmarks, as recorded by their coordinates; or (2) log-transformed straight-line distances between landmarks, taken from samples of organisms of varying sizes. Other types of data, e.g., meristic and qualitative variables, are best treated separately because they are not compatible with the deformation model. In the following sections we will discuss aspects of size variables (and their log-transformations) and aspects of the "adjustment" of shape for size, which follow from their origin in the geometry of the data.

2.2.1 Logarithmic transformations

Since Huxley's pioneering exposition of 1932, statistical studies of allometry have generally invoked all the common measures of extent—length, area, volume, weight—in terms of their logarithms. So strong is this presumption that the transformation is occasionally applied totally inappropriately, as to Cartesian coordinates (see Sec. 3.1.3). We have found at least five justifications for the resort to log-transformed distances as the basic variables of multivariate morphometric modeling. We believe that a maneuver so common deserves more than the scanty arguments devoted to its exposition nowadays (e.g., Smith, 1980).

We consider the most persuasive rationale for log-transformation to be the argument from the factor model. **Allometry** is the study of the consequences of size for shape (Gould, 1966). In this context, "size" ought to be conceived as a **factor** or **latent variable** (Jöreskog and Wold, 1982) rather than any datum expressly measured (Secs. 1.2, 4.2.3). In this general class of models, observables have an expected value which depends linearly on the values of other variables observed or latent, with an error variance about that expectation. Whatever size S is to be, the dependence of our observed distances (or their transformations) on it will be that of the factor model, having the form

$$d_i = \mu_i + a_i S + e_i \qquad (2.2.1)$$

where d_i is the ith distance measure or transform, μ_i is a correction for the mean of d_i (see below), a_i is a positive "loading" or hypothetical regression coefficient, and e_i is the random error term. This seems to be the simplest statistical embodiment of our natural assumption that all empirical indicators of size should have trends of the same sign. (Such a model is easily altered to accept interspecific effects as well; see Section 4.3.) In an analysis of a single group, the first principal component of log-transformed data fits the requirements of this model. For a set of distance measures, this component is normally characterized by consistently positive loadings corresponding to a joint increase (or decrease) of all variables. Whenever growth is expressed as a joint upward trend of all variables, this first component is usually interpreted as a growth trend (Jolicoeur and Mosimann, 1960) or, in our terms, as a latent variable serving as a proxy for size. In this way loadings derived from distance measurements provide a description of growth as a transformation. Allometry is indicated by unequal loadings of variables on this first component, and biological interpretation of allometric data proceeds using the coefficients of these regressions.

In this model the log-transform corresponds to a certain special null hypothesis, namely, isometry. In the case of growth without change in shape—strict Euclidean similarity—we wish to see that all coefficients a_i indicate equivalence of response. For such perfectly isometric data, if the dependent variable of Equation 2.2.1 is the raw distance measure, then each distance will have a regression coefficient proportional to its own typical or mean value. We would like isometry to appear instead as a collection of coefficients that are equal; so we must correct each loading for that variable's own scale, by dividing out its mean.

This division results in a derived variable d_i/μ_i whose regression coefficient on S closely approximates the regression coefficient of log d_i upon this same predictor. The logarithm function is the integral of the function $1/x$, so that for x_1 and x_2 sufficiently near the mean μ_i, the quantity $(x_1 - x_2)/\mu_i$ approximates the first-order Taylor expansion of the quantity log x_1 − log x_2. (This is the same formal maneuver involved in the derivation of the differential form of Huxley's model, $dy/y = k\, dx/x$. The

2.2 The Derived Data of Size and Shape

allometric constant k of the equation $y = bx^k$ is the ratio of the two loadings a_y, a_x in the factor model.)

Isometry is a special case of the more general **allometric factor model**, in which the a_i are not constrained. The unique position of isometry as a null hypothesis strongly suggests the transformation to logarithms to make the factor loadings interpretable. Another common transform which would guarantee identity of all loadings under a slightly different hypothesis (a's *and* error variances equal) is the conversion from covariance to correlation matrices, scaling each variable to its own standard deviation (rather than its mean, which led to the logarithm). However, this tactic has a serious flaw: for data that precisely fit the allometric model we may get identical loadings a_i whether or not the data fit the isometric model. We believe this point was first explicitly presented by Jolicoeur (1963), although Tukey (1954) argued it earlier in the context of morphometric path analysis upon observed variables.

The model (Eqn. 2.2.1) without log transformations implies that as size increases indefinitely the ratio d_1/d_2 between any two dependent distance measures should approach an asymptote, the value a_1/a_2. Empirical ratios of raw size measures do not usually tend to approach asymptotes in this way, even ratios of quantities all measured in the same units. Instead most raw shape ratios are correlated with size in all parts of the size range. The unique "size variable" uncorrelated with all these loglinear "shape variables" (Mosimann, 1975) has no particular utility in the factor model (see Sec. 4.2.3).

A second justification for the log transformation is likewise prior to inspection of particular data: it is the argument from physical dimension. In the biophysical laws which regulate biological shape, lengths appear more often in products or powers than in sums—in expressions for surface area, volume, mechanical moments, and the like (Maynard Smith, 1968). The common shape measures fall into this same category, as ratios of lengths, that is, one length times the -1 power of another. If our models are linear in the logarithms of lengths then the logs of all these expressions are likewise linear and we can invoke that most convenient statistical machine, the vector algebra of linear modeling. We can even include measures in other units, such as weight, heat loss, or information, if they can be reduced to powers of millimeters by approximately constant conversion factors. If the underlying models are instead linear in the lengths, then the partial derivatives of the products are not constant over the natural ranges of the predictors, so that we cannot execute path analyses or even interpret individual regression coefficients without clumsy interaction terms. This consideration leads us to the use of length-elements instead of long measured distances wherever possible, in the method of trusses (Sec. 3.2) and the method of biorthogonal grids (Sec. 4.5).

Other arguments encountered in defense of the log transformation have a more ad hoc flavor than these. We notice empirically, for instance, that in homogeneous populations the standard deviations of morphometric variables are to a great extent proportional to their means. Dividing each variable by its standard deviation would destroy any information borne in the variation of standard deviation independent of scale, so that we would do much better (see Gnanadesikan, 1977) to attempt a power transformation with an exponent near zero. The log transformation may be viewed as lying between power transformations of small positive exponent and power transformations of small negative exponent.

Or we might notice that the standard deviations of morphometric variables conditioned on constant size (that is, the magnitudes of the terms e_i representing errors of allometric regulation) are likewise proportional to the means of the variables, suggesting that the log transformation be invoked in the interest of **homoscedasticity**.

Finally, one often notices that scatters of motley bivariate data, such as are involved in the study of intraspecific associations, appear to lie straighter after log-log plotting than before. The association should not be summarized by its correlation as in Smith (1980): in the context of allometry it is the covariance or the regression slope that is of greater interest. Such data could probably be fitted even better by power transformations to exponents not quite zero. The use of the log transform in particular must be motivated by other, more elegant concerns based in the larger analytic context: the comparison of loadings a_i among all the variables.

2.2.1.1 The use of covariance matrices.—For calculation of principal components among distance measures we recommend use of the variance-covariance matrix rather than the corresponding correlation matrix. The eigenvectors of the two matrices are not the same, nor is it possible to pass from one eigenanalysis to the other by simply scaling the coefficients. Many systematists use correlation matrices in principal component analysis, in keeping with the usage established by psychometricians. If the variables are indeed in different units of measurement, then linear combinations of the original quantities would have little meaning and the standardized variates and correlation matrix might be employed. But if units of measurement are comparable, as with morphometric data, then the covariance eigenvectors have the greater appeal for the following reasons: (1) the analysis may be carried out in the original (log-transformed) character space, leading to a direct interpretation of character loadings on components and a direct comparison between populations; (2) because the ith principal component is defined as that linear combination of variables that accounts for the ith largest portion of the total character variance, maximization of the total in standardized data may give undue weight to less variable, relatively less precise measurements; and (3) the sampling theory for eigenvalues and eigenvectors computed from correlation matrices is exceedingly more complex than that for covariance matrices (Anderson, 1963).

2.2.2 Two comments on size adjustment

We work with size and shape mainly in terms of their changes, which are *transformations*, that is to say, comparisons among forms. Whether changes are expressed as explicit homology maps or as factors of measured distances, particular empirical comparisons will have an aspect of size-change together with one of shape-

2.2 The Derived Data of Size and Shape

change. In Section 4.3, for instance, we show how to extract these two aspects jointly as factors, and in Section 4.7 we extract them, again jointly, as components of the biorthogonal grid. Our treatment of size and shape as commensurate differs greatly from customary morphometric procedures. Before proceeding with our exposition, we comment on the usual fate of size information in morphometrics: size is computed specifically in order to be separated from shape information, then disregarded. This custom leads to inefficiencies and paradoxes in the analysis of many common sorts of distance data.

2.2.2.1 Why size adjustment cannot adjust for size.—We will claim throughout this book that, except in the concise presentation of conclusions, the usual size adjustment procedures (restricting size classes, ratios, regression residuals) have little if any useful role in most morphometric analyses.† The purpose of passing to ratios or residuals as measures or restricting the size-range of specimens is usually to remove from observed covariation the "confounding effects" of size. In our view, size ought not to be removed from observed measures. It does not explain somehow-irrelevant variance but rather perfectly meaningful *co*variance. In all statistical manipulations it should remain as one joint cause among several (species, sex, environment, etc.) of the associations empirically observed.

Restricting group comparisons to age or length classes (e.g., two-year-olds or adults between 50 and 65 mm in length) disregards ontogenetic shape-change within groups. We find this information necessary for meaningful descriptions of group differences. Furthermore, such restricted comparisons are not independent of size if any range of sizes is present or if the groups differ in mean size.

Size adjustment procedures usually result in a set of "size-free shape variables" which are then used to explore shape differences among groups. We shall demonstrate that this approach is usually misleading. Unless the effects of size and group membership on the original measured variables are known a priori, the resulting size adjustment is flawed conceptually, and often computationally. Size-adjusted variables can yield satisfactory ordinations but not biological explanation in terms of the original variables. It is one of our tenets that morphometric analyses must treat size-change and shape-change conjointly, so that the coefficients that result are interpretable.

That we should not remove size from measured variables follows from a theorem of Mosimann (1970) (more accessibly presented in Mosimann and James, 1979). In general, ratios or residuals are still statistically dependent upon all measures of size except at most a single one not computable a priori. The clearest intuitive approach to this theorem is geometric, as follows.

As usual, suppose we have measured a sample of organisms by homologous distance variables $X_1, X_2, \ldots, X_k$. The logarithms of the X's comprise a new set of

---

† We are not, of course, proscribing size *standardization*, as by way of predicted regression estimates. See Section 4.4.

variables $Y_1, \ldots, Y_k$. The ratios we are criticizing have the variables $Y_i - Y_j$ as their logarithms; the residuals will be taken among the Y's directly.

Consider, more generally, the set of all linear combinations $a_1 Y_1 + \ldots + a_k Y_k$ of these logarithms, the new variable $\Sigma a_i Y_i$. This is the logarithm of the product $X_1^{a_1} X_2^{a_2} \ldots X_k^{a_k}$ of powers of the original distances. If we multiply all the X's by a constant factor s (i.e., enlarge the form s times, or diminish our measuring unit by a factor of s) the product is multiplied by $s^{[a_1 + \ldots + a_k]}$. If the sum of the coefficients a_i is 1, this product of the X's scales linearly in s to serve as another size measure; $\Sigma a_i Y_i$ is its logarithm. In this section we call these products $X_1^{a_1} \ldots X_k^{a_k}$ with $\Sigma a_i = 1$ the set of **size variables**, and the $\Sigma a_i Y_i$ the **log size variables**.

All the log size variables may be diagrammed in what is known as their **dual space**, which we shall review here briefly. A thorough exposition may be found in Dempster (1969).

It is always possible to visualize the variables $Y_1, \ldots, Y_k$ as a set of vectors, also called $Y_1, \ldots, Y_k$, originating at a single point O, such that (1) the squared length of the vector Y_i equals the variance of the variable Y_i, and (2) the cosine of the angle between the vectors Y_i and Y_j equals their correlation as variables. These vectors can be imagined to lie in Euclidean space of at most k dimensions. For instance, Figure 2.2.1 sketches some three-dimensional configurations corresponding to the variance-covariance matrices indicated. The configurations can be rotated any which way in space—it is only the relative orientation of the vectors that matters.

In this Euclidean space of the diagram we may use the vectors $Y_1, \ldots, Y_k$ to assign coordinates to all the other points (i.e., as a **basis**). Any point $(a_1, \ldots, a_k)$ is the vector sum of the vectors $a_1 Y_1, \ldots, a_k Y_k$. One obtains it by traveling from O a distance along Y_1 equal to a_1 times the length (standard deviation) of Y_1, then from the end of Y_1 parallel to Y_2 a distance equal to a_2 times the length of Y_2, and so forth. By this formalism the corroboration of variance by squared length and of correlation by cosine is extended to all **linear combinations** of the original variables. The squared distance of the point $(a_1, \ldots, a_k)$ from O will exactly equal the variance of the variable $a_1 Y_1 + \ldots + a_k Y_k$, and the correlation between two linear combinations $a_1 Y_1 + \ldots$ and $b_1 Y_1 + \ldots$ will exactly match the cosine of the angle between the vectors joining O to the points $(a_1, \ldots)$ and $(b_1, \ldots)$. In particular, if the two linear combinations are uncorrelated, then the two vectors corresponding are perpendicular.

In the dual space we have been drawing, the combinations $\Sigma a_i Y_i$ with $\Sigma a_i = 1$—the logs of size variables—lie on a **hyperplane**, a flat Euclidean subspace of dimension one less than the dimension of the space in which it is embedded. For the case $k = 3$ of the figures here, the set of all size variables has dimension $3 - 1 = 2$—it is an ordinary plane. Each original log variable Y_i is in this plane, as its coefficients comprise a single 1 and the rest 0's, totaling 1, as required: $Y_i = 0 \cdot Y_1 + \ldots + 1 \cdot Y_i + \ldots + 0 \cdot Y_k$. The plane of logs of size variables is then the unique plane through the tips of the vectors Y_1, Y_2, Y_3, drawn as P_Z in Figure 2.2.2.

Logs of the usual ratios X_i/X_j are combinations $Y_i - Y_j$, whose coefficients in the Y's sum to zero. These are instances of the general linear combination $a_1 Y_1 + \ldots +$

2.2 The Derived Data of Size and Shape

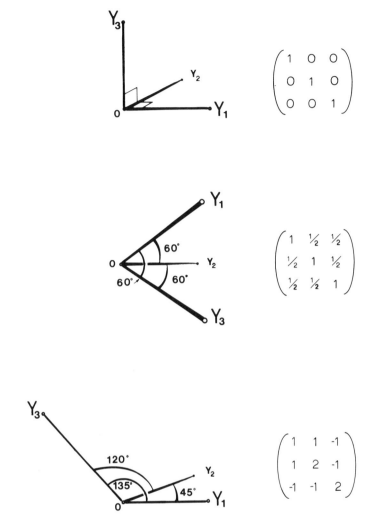

Figure 2.2.1 Schematic of dual space for three variables. Three triads of vectors, drawn in perspective, and the covariance matrices for the variables Y_1, Y_2, Y_3 to which they correspond. See text.

$a_k Y_k$ whose a's sum to zero. In the dual space, such variables lie in another hyperplane, drawn as P_H in Figure 2.2.3. Each vector in this plane is the difference of two vectors of P_Z in many different ways. Because the ratios of which the points of P_H are the logarithms scale to dimension zero in length, we call them **shape variables**, and the corresponding $\Sigma a_i Y_i$ the **log shape variables**. Then P_H is the plane of logs of shape variables, in our restricted sense, just as P_Z is the plane of logs of size variables.

Because no set of coefficients can sum to both zero and one, the planes P_H and P_Z have no points in common: they are geometrically *parallel*, and have been so drawn. P_H may be characterized as the unique plane through O parallel to the log size plane P_Z.

2 FOUNDATIONS

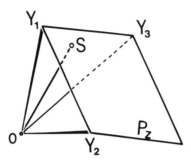

Figure 2.2.2 The plane of size vectors. All size vectors, such as S, end on the plane P_Z through the endpoints of the observed size variables Y_1, Y_2, Y_3 interpreted as vectors.

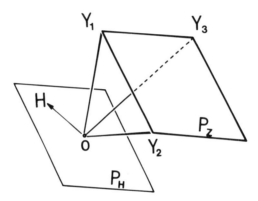

Figure 2.2.3 The plane of shape vectors. All shape vectors, such as H, lie in the plane P_H parallel to P_Z through O.

Mosimann's central theorem states that at most one size variable is independent of the entire space of shape variables.† In a geometric diagram, "independence" is read as orthogonality. In our terms Mosimann's theorem therefore implies the following assertion: just one log size variable may be orthogonal to the whole plane of log shape variables. That is, only one vector from O to P_Z may be perpendicular to the plane P_H.

But this is geometrically obvious. Precisely one normal to a plane may be drawn through any point of space. In other words, there is a unique vector $\tilde{S}$ (Fig. 2.2.4) from O normal to P_Z and therefore, because the planes are parallel, normal to P_H as well. As a variable, $\tilde{S}$ is orthogonal to all the variables in P_H, that is, all the log shape variables. In this context it is the unique size variable of Mosimann's theorem. It is unlikely, of course, that $\tilde{S}$ is identical with any of the Y's, i.e., that it was measured.

---

† Mosimann's publications treat a more general class of size and shape variables than the linear vector space to which they are restricted here. Our Figure 2.2.4 certainly does not constitute a proof of the theorem in that more general context.

2.2 The Derived Data of Size and Shape

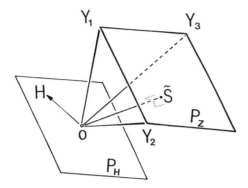

Figure 2.2.4 The size vector $\tilde{S}$. There is a unique size variable $\tilde{S}$ orthogonal to (i.e., uncorrelated with) all shape variables. It is represented by the vector through O normal to the size plane P_Z.

We can compute the coordinates $(\tilde{a}_1, \ldots, \tilde{a}_k)$ of $\tilde{S}$ from the covariance matrix (σ_{ij}) among the Y's. Each variable $Y_i - Y_j$ must be uncorrelated with $\tilde{S}$; then, as

$$\operatorname{cov}(\tilde{S}, Y_i - Y_j) = \operatorname{cov}(\tilde{S}, Y_i) - \operatorname{cov}(\tilde{S}, Y_j),$$

$\tilde{S}$ must have equal covariances with each of the Y's:

$$\operatorname{cov}(\Sigma \tilde{a}_i Y_i, Y_j) = \text{const.}$$

or, for each j,

$$\Sigma \tilde{a}_i \sigma_{ij} = \text{const.}$$

In terms of a vector **1** of k 1's, we may solve for $\tilde{S}$:

$$\tilde{S} = \text{const.} \times \mathbf{1}\, (\sigma_{ij})^{-1},$$

so that

$$\tilde{a}_i = \text{const.} \times \Sigma_j \tau_{ij}$$

where (τ_{ij}) is the matrix inverse to (σ_{ij}). The constant is $(\Sigma\Sigma_{i,j}\tau_{ij})^{-1}$.

The only sense in which "size" can be partialled out via conversion to ratios is size as measured by $\tilde{S}$. For any other log size measure, the set of log shape measures to which it is orthogonal is represented by a line in P_H (in general, the intersection of two hyperplanes), upon which the vast majority of ratios will *not* lie. Likewise, for any log shape variable H there is merely a line of log size variables with which it is uncorrelated—the line L through $\tilde{S}$ perpendicular to the line from $\tilde{S}$ through $\tilde{S} + H$ in the plane P_Z (Fig. 2.2.5). Because (1) only one log size measure $\tilde{S}$ is generally corrected for by the use of ratios, and (2) different ratios correct for different sets of log size measures, the purpose of taking ratios is fundamentally unachievable.

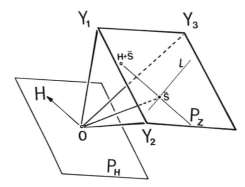

Figure 2.2.5 Correlations between shape and size variables. For any shape variable H, consider the vector $H + \tilde{S}$ in the plane P_Z and the line L through $\tilde{S}$ perpendicular to the line from $\tilde{S}$ to $H + \tilde{S}$ in P_Z. The shape variable H is uncorrelated with size variables only when they lie upon L. The size variable $\tilde{S}$ is on all such lines because it is orthogonal to all shape variables.

It can easily be the case that certain log shape variables correlate positively with all the observed log size variables (Fig. 2.2.6), a circumstance seriously conflicting with the import of size and shape as conceptually orthogonal. This will happen whenever one of the coefficients $\tilde{a}_i$ of $\tilde{S}$ is negative, which can be shown to be equivalent to the regression coefficients of one of the Y's upon the others summing to more than 1. In practice this criterion must often be unwittingly fulfilled, as for instance in cases of diverse coefficients of size allometry or of high **collinearity** among the variables. When a ratio of distances is positively correlated with all size variables, including its own denominator, it is clearly pointless to speak about "correcting shape for size" by the taking of ratios (Mosimann, 1970; Atchley et al., 1976).

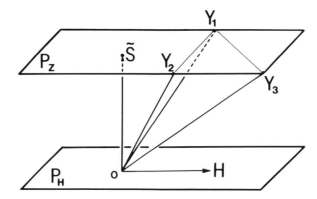

Figure 2.2.6 Unwanted correlations between shape and size. For some patterns of covariance among the size variables, there exist shape variables like this H that are positively correlated with all the measured size variables Y_1, Y_2, Y_3.

2.2 The Derived Data of Size and Shape

Like ratios as "shape variables," residuals depend upon the choice of a size variable. Different size variables may substantially alter the covariance structure of the residuals, and hence alter the coefficients of subsequent multivariate analyses. In addition, the robustness of residuals as shape variables depends upon the size distributions of the groups, the within-group slopes and the choice of the regression model.

Three types of regression models have been used to generate new variables. The **pooled among-group regression** will almost always generate size-dependent residuals (Fig. 2.2.7a). The **pooled within-group regression** will produce robust residuals for shape analysis if and only if the groups have identical size distributions and parallel within-group slopes for the particular size variable chosen (Fig. 2.2.7b). **Separate within-group regressions** (Fig. 2.2.7c) may be used to generate residuals, if the pooled within-group model is inappropriate, but the group means must be restored to the "residuals." This latter procedure does not produce residuals in the usual sense; subsequent factor analyses will recover the group differences only because they have been restored at this stage.

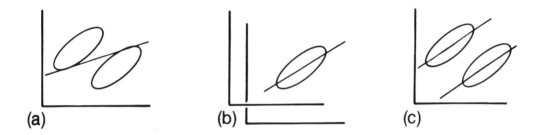

Figure 2.2.7 Three regression models for size-correction. (a) Pooled among-group regression. (b) Pooled within-group regression. (c) Separate within-group regressions.

All the conclusions regarding ratio-adjustment may be applied to residuals as well. The measured variables play the role of the Y's directly. In the dual space of our diagrams, the construction of shape variables as residuals may be drawn (Fig. 2.2.8) as projections onto the plane through O normal to the independent variable. If size is measured by $\tilde{S}$, the space of its regression residuals is just the plane P_H drawn before. If size is measured instead as any other variable, e.g., one of the measured Y's, or their sum, or their principal component Y^*, then the space of regression residuals may be fairly well-determined geometrically—it will probably be very near the plane drawn here.

Notice, for collinear data, how sensitive is the detailed geometry of these residuals to the precise choice of independent variable, i.e., normal vector. Slight differences in the choice of a regressor size variable can result in great changes in the values of the residuals, their contributions to subsequent discriminant functions, and so on. From

34   2 FOUNDATIONS

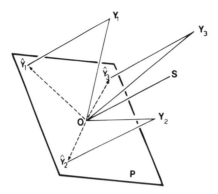

Figure 2.2.8 Construction of shape variables as residuals. In ordinary practice, shape variables are generated from observed size measures by regressing out one particular size measure. In dual space this is the projection of the size measures onto a plane P through O normal to the size vector S chosen as regressor.

Figure 2.2.9 it can be seen how very small changes in the direction of the independent variable result in profound reallocations of variance among the residuals $\hat{Y}_i$. In particular, the first principal component Y^* is computed so that the residuals have least summed squared variance, that is, so that the arrows from O to $\hat{Y}_1, \ldots, \hat{Y}_k$ have least summed squared length. This criterion does not refer to their angles (covariances) in any way; so the correlations observed among the $\hat{Y}$'s are partly artifacts of that optimization.

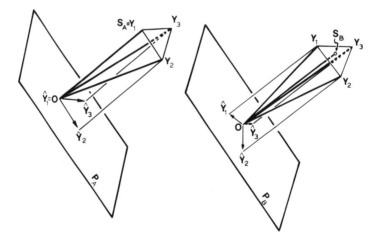

Figure 2.2.9 Instability of covariances among residuals. For highly collinear size variables, small adjustments in the choice of a regressor size variable (S_A or S_B) result in substantial changes in the covariances observed among the residual shape variables. Plane P_A is perpendicular to size variable $S_A = Y_1$, and plane P_B to size vector $S_B = \Sigma Y_i / 3$.

2.2 The Derived Data of Size and Shape

As the size variable changes, so do the variances and covariances of the residuals. Consequently, coefficients from their subsequent multivariate analyses are unstable even if the ordination of taxa is somehow stable. Here we encounter a paradox: the dependency of residuals (shape variables) upon the choice of a size variable increases with the collinearity of the original variables. That is, the better size is characterized, the noisier are all shape residuals.

In Section 4.2 we will explain a technique, Sewall Wright's factor analysis, for which these covariances *are* meaningful in the context of models invoking multiple explanatory factors, not merely "size." In these models size is not an adjustment measured as $\tilde{S}$ or otherwise, and the other explanatory variables explain more than "residuals from size." Whenever size is used instead as an isolated "correction," the resulting adjustment is in conflict with later explanations for which the shape variables are themselves used as predictors.

We shall not refer again to the notions of size and shape variables and log size and log shape variables as defined in this section. We will define "size" and "shape" variables in the context of particular models and particular covariance matrices, and they will not be restricted to hyperplanes of constant Σa_i or otherwise. When a shape factor is computed in the light of a particular size factor, both will be present in all ensuing statistical manipulations and explanations.

2.2.2.2 The Kluge-Kerfoot "phenomenon": artifact of size correction.—The inability of ratios to effect automatic size "adjustment" is illustrated rather ironically by the familiar **Kluge-Kerfoot phenomenon**, the apparent correlation *over characters* of amount of interlocality differentiation with amount of intrapopulation variance (Kluge and Kerfoot, 1973). For such a correlation to be computed one must make commensurate the scales of measurement of the characters making up a suite. Rohlf et al. (1983:198) have shown that the conventional adjustment, the scaling of all variables to mean 1, supplies the typically positive Kluge-Kerfoot correlations as mere artifacts of the division: "The observed correlation is due to the measures of within- and among-population variability both being functions of a third variable—the sample mean."

The use of the grand mean in this discussion is incidental. Scaling to any criterion will induce exactly the same sort of correlation. In fact it is impossible to construe a scaled Kluge-Kerfoot correlation meaningfully: there is not enough information at hand to set up a pair of vectors between which a correlation can be computed. The situation is the two-dimensional version of the preceding discussion: in the context of only two size variables, there can be only one shape variable, which cannot be compared with others. Character by character, we have only two measures of variability, that within (W) and that between (B). If one or the other, or some function of the two, or some proxy (such as range), must be used up in setting a scale for comparison with the variabilities of other characters, then *there remains only one degree of freedom per character*, one statistic only, expressed perhaps as W/B. As there are in practice no remaining aspects of morphometric characters with which to correlate this ratio, we are reduced to a one-by-one covariance matrix, which is of little interest. For instance, we would certainly not consider the sample mean in this context, because it is as uncorrected for scale as the measures of variance with which we began. Dividing by a measure uncorrected for scale does not correct for scale.

Properly construed, there is no artifact in the usual correlation of "adjusted W" with "adjusted B" because there is no correlation to be had. Rohlf et al. (1983:184-5) discover this when they note that in their models A and B (morphometric variables normal or log-normal with intrapopulation coefficients of variation the same for all populations) the scaled variance Between is an exact linear function of the scaled variation Within, so that they are left with only one measure to vary over characters; but they fail to appreciate the import of this fact.

One cannot adjust for size in the absence of a model for the effect of size on the data: this is as true of "size of characters" as of "size of organisms." The Kluge-Kerfoot literature proposes no models of the dependence of W or B on size over characters, and so size cannot be partialled out, by sample mean or by any other measurable quantity, while still leaving a pair of measures.

For instance, the common one-factor model of size allometry models each character y_i as a regression with error:

$$y_i = a_i S + e_i$$

where S is Size, a "general factor" (see Sec. 4.2) and e_i is a random error with variance σ_i^2, uncorrelated with S and with all the other e_j. If we have populations for each of which S has mean $\bar{S}_n$, $n = 1, \ldots, N$, and intrapopulation variance σ_S^2 independent of n, then within populations

$$\text{var}(y_i) = a_i^2 \sigma_S^2 + \sigma_i^2$$

while between populations

$$\text{var}(y_i) = a_i^2 \, \text{var}_n(\bar{S}_n) \ .$$

These certainly look confounded, and with the correct sign—a constant positive multiple of a_i^2 appears in both expressions. We would be tempted, in fact, to refer to a_i^2 and σ_i^2 as the "proper," unconfounded estimates of B and W. But we have not set any scales. *Nor can we*, meaningfully. Suppose, for instance, we scale the intrapopulation variance of each y_i to 1. Then a_i^2 and σ_i^2 are forced into a perfect negative correlation, frustrating our goal of examining the correlation as an empirical phenomenon.

It is our belief that all models of character covariance for which scale is not fixed a priori (i.e., centimeters, or log centimeters) will be scale-dependent in this way and will therefore make impossible the Kluge-Kerfoot computation in any legitimate fashion. In short, characters cannot be made commensurate without artifact, unless they are commensurate to begin with, such as the geometric networks of distances and directions that are the subject of this book.

2.3 Why Not to Omit Homology Information

Morphometrics, in our view, is the statistical analysis of biological homology treated as a geometrical deformation. Its two sources of information, geometric location and biological homology, should be equally important aspects of any morphometric analysis. We believe that in most contexts all variables should be based on distances measured homologously from form to form, either distances among landmarks declared to be homologous at the outset or lengths of line-elements homologous according to a formal computation beginning with landmarks. When there is available information about the curving of form between landmarks, we suggest (see Sec. 3.3.2) that these locations not be construed as variables directly but instead be combined into an archive of the form. Pairs of archives will eventually be converted (Sec. 4.5) into a computed homology map from which relations among homologous distances can be extracted directly.

Hence the discipline of measuring only biologically homologous quantities often forces the postponement of constructing explicit "variables" to the last phase of an analysis. A custom has arisen of circumventing such rigor by measuring the curving of single forms without regard for homology. The quantities that result are then submitted to multivariate statistical analysis just as if they were homologously measured variables. But it seems pointless to us to execute optimal statistical computations, such as Fisherian discrimination, when up to half the information of the data base has been intentionally discarded in the course of measurement. We find the purely geometrical approach unsatisfactory whenever there is *any* information about biological homology present—any landmarks at all, or any evidence for allometry.

It may sometimes appear that one can execute a satisfactory discrimination in terms of these coefficients, however flawed, insofar as some of the information separating natural groups ought to be encoded in outline shape regardless of whether landmarks were designated. Our criticism is not that systematic differences in form are ignored by the functional methods but that they are made uninterpretable. In this section we demonstrate the difficulty of restoring the missing information coherently in the course of any subsequent analysis.

2.3.1 Forms as geometrical functions

Biological outlines that are drawn as closed curves may be modeled as periodic functions, points cycling around and around. Often the cycle may be quantified by a **radius function**, the distance to the curve from a center (suitably chosen) as a function of angle with respect to a starting ray (suitably chosen).† For data of special forms

---

† For sufficiently smooth data, the dependence on a center may be eliminated by recourse to the **intrinsic representation** of a curve, its curvature as a function of arc-length measured, once again, from a starting point suitably chosen.

other sorts of functions may be invoked. For instance, outlines symmetric about a straight line may be described by a **width function**: perpendicular distance from that line as a function of position along it.

Corresponding to each of these geometric domains there is a mathematically (but only mathematically) natural decomposition of empirical functions into a unique sum of **orthogonal components**: the **orthogonal functional analysis** or **orthogonal decomposition**. The exact form of the components is determined by the domain: sines and cosines for periodic functions, the Legendre polynomials for functions on a line, and the spherical harmonics for distances from a point in space. The coefficient corresponding to each component is its covariance with the empirical function, computed by integration around the form.† Increasing numbers of systematists and biometricians treat the coefficients generated by these decompositions, usually the Fourier amplitude analysis of a radius function, as a set of meaningful shape descriptors used for variables in discriminatory or other multivariate analyses (Anstey and Delmet, 1973; Younker and Ehrlich, 1977, and references therein).

We believe that in most morphometric applications this class of practices is inadequate. There are more data in an outline than the mere curve of its points in two dimensions; the homology function by which we describe correspondences between outlines is an independent source of biometric information. (See Bookstein, 1978a: chapter iv.) In Figure 2.3.1a, for instance, the Fourier method, having no way of coping with the black dots, would consider the forms to be identical. But if the dots were homologues, the forms would differ by a biological transformation of considerable significance.

The fundamental consequence of this loss of data for the purely geometrical methods is as sketched in Figure 2.3.1b. Suppose we notice a difference between two groups, or a correlation with some cline, in the value of a numerical function (e.g., width away from an axis, or radius from a center) evaluated "at" some point of an outline. The reason reference to position is qualified is that without access to a homology function we cannot verify that the function was evaluated at corresponding loci in the two forms. For this reason we cannot differentiate between increase or decrease in the value of the function (arrow F in Figure 2.3.1b) and a shift of the point at which we are evaluating it (arrow H). Each of these effects is biologically meaningful, but they represent, so to speak, orthogonal modes of description.‡ They are inextricably confounded in the one numerical difference.

---

† Another class of techniques treats an entire image, interior together with boundary, as a function in two dimensions whose value at a point is the pictorial darkness there. Because the Fourier transform of an image encodes our notion of repetitive spatial pattern very well, decomposition of images in this way is a valid means of searching for spatial texture (e.g., Oxnard, 1980). The general criticisms of this section do not apply in this case.
‡ This distinction corresponds to the difference between fixed and co-moving derivatives in continuum mechanics. See Skalak et al. (1982).

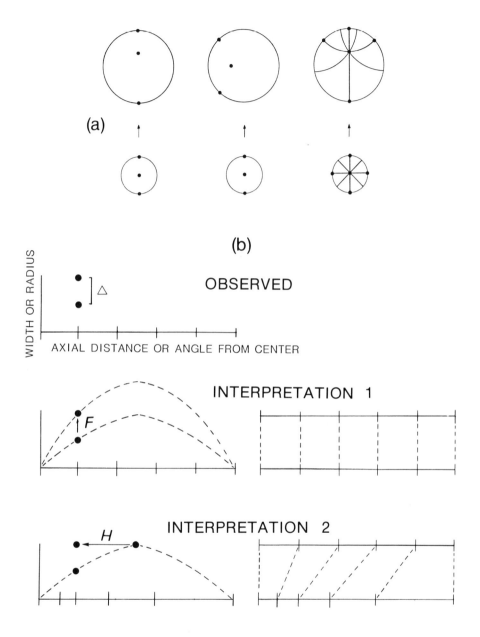

Figure 2.3.1 The ambiguity of homology-free analysis of outlines. (a) A significant transformation of form leaving landmark-free "shape" and Fourier coefficients unchanged. (From Bookstein, 1978a: fig. V-1.) (b) When a function "at" a particular locus upon an outline appears to change its value, we need information about homology to discriminate between a change in the value of the function (F) and a change in the homology of the locus (H). The two modes of description are perpendicular upon this diagram.

40   2 FOUNDATIONS

The techniques of orthogonal decomposition, such as Fourier analysis, compound this difficulty by integrating the ambiguity all the way around a circle of directions or from end to end of an axis. After the integration, change in a component could be expressing any combination of these two orthogonal modes of description anywhere in the domain of the integral. This loss of information is particularly serious regarding local features of form. The easiest description of a bulge, for instance, is hierarchical: its position relative to the form as a whole, together with partial descriptors of its specific shape. The purely geometric method confuses change in shape of the bulge with change in its position; the additional maneuver of orthogonal decomposition, which is necessarily not local but global, blurs these two alternatives further among indefinitely many general "components." The simpler underlying descriptors cannot be recovered except by reconstructing the form from a very large number of coefficients, then quantifying it again from the beginning.

Because relative spacing of landmarks around the outline is ignored, and size is usually expressed in this relative spacing, there is no way to partial size effects out of an orthogonal functional analysis. One cannot standardize the descriptor vector by any linear maneuver. Although the zero-degree component usually measures some aspect of width directly, the allometries (relations of shape and size) are confounded by the method. Partly they appear in the correlations between the coefficient of the zeroth component and subsequent coefficients—although these correlations are necessarily zero for the single form, they need not be zero over the forms of a sample. All the rest of allometry is obscured in the repositioning of landmarks and in the dependence of standardization and centering upon size.

2.3.2 The confounding of homology in purely geometrical analysis: Examples

Stoermer and Ladewski (1982) modeled the shapes of certain pennate diatoms (Fig. 2.3.2a) using two landmarks and two arcs. The landmarks were set at the ends of the form in the valve view. (The head end, more laterally expanded, was placed on the right.) These ends were connected by one arc above and its mirror image below. The length (diameter) of the valvar axis was then standardized to 1.0 and its width transcribed as a function of position along the axis joining the landmarks (Fig. 2.3.2b). There is a third landmark, the stigma, visible in the specimens of this analysis along the axis of each form near the middle of its length and likewise near, but not precisely at, the widest point of the profile, corresponding to the central nodule.

From direct multivariate analysis of this width function (a strategy similar to Lohmann, 1983) there result principal components of shape variation that may be visualized by their effects on a typical valve profile.† One component in particular, having the effect shown in Figure 2.3.2c, seems to represent an interesting decrease of

---

† The effects are visualized by constructing outlines having varying scores on this component and fixed scores on all the other principal components. We use the same method in Section 5.4.

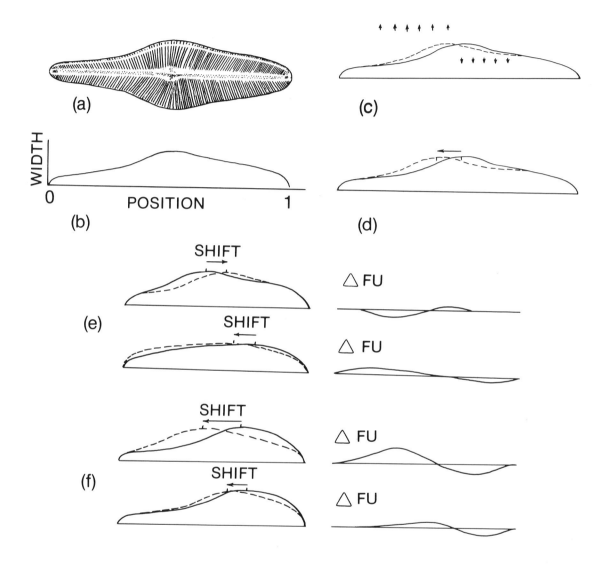

Figure 2.3.2 Critique of the geometric analysis of shape variation. (a) Profile of a typical pennate diatom of the genus *Gomphoneis*. (b) Representing a standardized form by its width function. (c) The second principal component of shape variation, visualized as a change in the width function. (d) The same component viewed more simply as a change in biological homology along the axis of the form. (e) Applied to diatoms of different shapes, a fixed shift of axial homology yields different geometrical "components." (f) The shift "component" is not linearly related to the extent of the shift. Doubling the shift axially does not double the effect on the width function. (Data after Stoermer and Ladewski, 1982.)

width near the middle together with an increase toward one end. Such a description is more obscure than it needs to be: vertical, in the metaphor of Figure 2.3.1b, when it might be horizontal. This component could as well be expressing a simple adjustment of homology: variation in the placement of the central nodule, Figure 2.3.2d. Such variation would have been measured better by a straightforward comparison of distances to the two ends of this projected form, just as scale was standardized according to the distance between the ends—but we cannot tell whether the component is expressing this horizontal variation or not!

This example typifies how purely geometrical representations of form must obscure adjustments of homology along the axis or around the circle over which the width or radius function is taken. Even if the component in Figure 2.3.2c,d were measuring something akin to a shift of the medial landmark, it does not behave as we would expect a measure of that landmark shift to behave. Shifts of the central nodule in diatoms of different mean shapes lead to different profiles on this component (see Fig. 2.3.2e); and variation in the extent of the shift is not closely associated with any linear variation of the width function (Fig. 2.3.2f).

We cannot tell to what extent a component that looks like this one is expressing an actual shift of homology (driven, perhaps, by changes in the ability of the organism to extend its axis toward the ends) and what part is change in width wholly independent of that shift—the necessary information is just not present in the data base. In a vocabulary lacking reference to *biological* changes in the domain of the *geometric* function, interpretation becomes much more difficult for multivariate components that partially express these variations of landmark configuration. This leads, in turn, to difficulties for interpreting subsequent components forced to be orthogonal to these and discriminant functions adjusted for such systematic variations within group.

A similar correlation between a purely geometric component (this one a Fourier amplitude) and the relative repositioning of one landmark is reported by Ehrlich et al. (1983) for certain planktonic foraminiferans. This correlation makes it impossible to interpret subsequent quantities in the analysis. The problem of homology along an axis is replaced by two: the problem of locating a homologous "center" at which to measure angles, and the possibility of true change in angles subtended by landmarks out of that center. Whenever there are three landmarks to be had along an axis, or four around a circuit, one cannot avoid the impact of landmark repositioning upon the domain of the geometric functional representation (Bookstein, 1983). The geometric style of description of any interesting trend will almost always confound the two sorts of interpretation of Figure 2.3.1b: change in the value of the function and change in the biological identity of the ostensibly fixed loci at which it is evaluated.

In the examples mentioned here, this confounding, perhaps owing to special symmetries, seems to devolve upon a single geometric coefficient, the import of which can be reinterpreted using the vocabulary of the homology function. But in most cases (and probably in these as well) the simple relative movement of a landmark is distributed over a great many components of the geometric representation. Elsewhere

2.3 Why Not to Omit Homology Information

(Bookstein et al., 1982) we present an extended demonstration of this difficulty in the context of Fourier analysis. Throughout these examples, the geometric analysis, even when it can later be coupled with homology information, is not measuring the repositioning of landmarks in a manner likely to lead us to simple parameters underlying the regulation of form. These confoundings of geometric coefficients with homology shifts in the domain of the function will normally resist detection after the fact; instead, one must establish data bases with this problem in mind at all times. This requires that landmarks actually be located in the course of digitizing, in which case further computation according to the geometric representation is pointless. One should instead use a method that takes into account both geometric location and biological homology, as recommended throughout this treatise.

3

Data Collection and Preparation

3.1 Patterns of Distance Measurements

Customarily, morphometric data are taken without regard for allometry or its variations among populations or growth stages. It has been common to study growth, for example, by analysis of body length or height or weight only; to describe the shapes of bones by measures of their lengths and widths; to characterize entire forms by relative lengths and breadths of head, trunk, tail, and appendages. Although such measures are deeply entrenched in the methodology of systematics, their usefulness in solving real biological problems may be limited. There are far more homologous measures on biological forms than are used in typical multivariate data sets, and results of morphometric analyses will depend upon the particular set of measurements chosen. If the selection of distance measures does not correspond by accident or design to the principal directions of form difference, the resulting descriptions of the differences between forms will be inadequate.

3.1.1 Conventional distance data sets

There are several biases and weaknesses inherent in traditional character sets such as those in Figure 3.1.1. They may be summarized as follows.

1. Most characters tend to be aligned with one of a very few axes, such as the "longitudinal," with only scant sampling of depth and breadth. Thus a large amount of information in the data is repetitious while other information, variation in oblique directions, is lacking.

2. Coverage of the form is highly uneven by region as well: dense in some areas of the body and sparse in others.

3. Some morphological landmarks, such as the tip of the snout and the posterior end of the vertebral column, are used repeatedly. Any uncertainty in the positions of these morphological features will be propagated through series of measurements.

4. Many landmarks are "extremal" rather than "anatomical" (sensu Moyers and Bookstein, 1979; Sec. 1.2.1). Anatomical landmarks are true homologous points identified by some consistent feature of the local morphology. Extremal landmarks are defined in terms of minimum or maximum distances (e.g., greatest body depth), and therefore their placement may not be homologous from form to form.

5. Many measurements are too long, traversing much of the body. Short distances contain more localized information than long ones; the longer the distance, the more its covariances ignore variability among its unmeasured subsegments. Thus, short segments are to be preferred to long ones, because the information borne in their loadings is better localized; in particular, longer sums of short vectors should be omitted.

The loadings resulting from our factor models will generally be regression coefficients of log segment-length on size or on shape-difference. There will be two such items of information, corresponding to directional data, only for points through which pass two segments of the sample (Fig. 3.1.2a). Were these distances selected optimally, the results would bear a direct relationship to biorthogonal analysis (Sec. 4.5). Yet from loadings in two directions, which is all one can expect through a typical point, one can derive no information whatever about the loadings in other directions (Fig. 3.1.2b) nor a comprehension of smooth directional variation. In principle, the data should supply loadings for many directions through each point of a closely spaced set (Fig. 3.1.2c). A suitable discrete approximation (Fig. 3.1.2d) would have multiple intersections and evenness of areal coverage.

Typical morphometric data are a most uneven sampling in comparison with what is possible. Multivariate morphometrics offers no system for selecting characters to be studied other than that they be numerous and sample the whole form. Consequently, traditional data sets often turn out to be highly biased in many of the ways described above, and success in selecting effective characters has been largely a matter of chance. Early in a study (and by early we mean prior to or during multivariate statistical analysis) it is fruitless to measure ratios and angles arbitrarily arranged about the form, because with only a small number of interpoint distance measures one can archive the entire configuration of landmarks and thus preserve *all* the information it contains. We do not measure "size" or "shape" of point-configurations; we simply transcribe them and, much later on, *derive* appropriate measures of form change or form comparison from analyses of the configurations as wholes.

3.1.2 Triangulations

The least redundant archives of distances are the **triangulations**. In these schemes the points are assigned an arbitrary order; all three distances are measured among the

3.1 Patterns of Distance Measurements 47

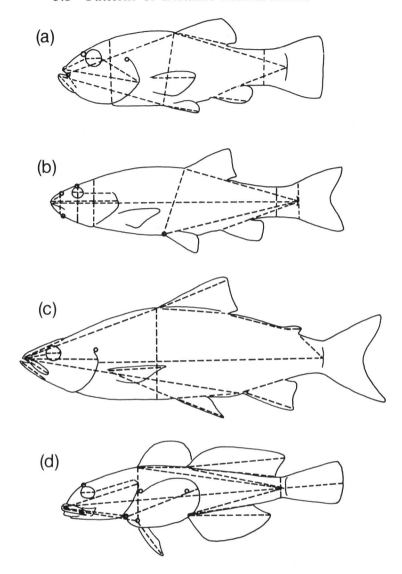

Figure 3.1.1 Four examples of more-or-less conventional sets of distance measures used in fish morphometrics. Dotted lines represent projections of the measurements onto the midsagittal plane. Small circles indicate positions of width measurements. (a) *Cyprinodon*; (b) minnows; (c) ciscoes; (d) *Cottus*.

first three points, and then from each subsequent point distances are measured to two points previously located (Fig. 3.1.3). Some of these schemes have a certain aesthetic appeal by reason of symmetry. For instance, we may begin with a set of landmarks taken without regard for the boundary curvature, as in Figure 3.1.4: a set of isolated loci no longer having any intrinsic order. Suppose there are n landmarks; here $n = 7$. In the form of Figure 3.1.4a each point is connected to its neighbors along the outline

48   **3 DATA COLLECTION AND PREPARATION**

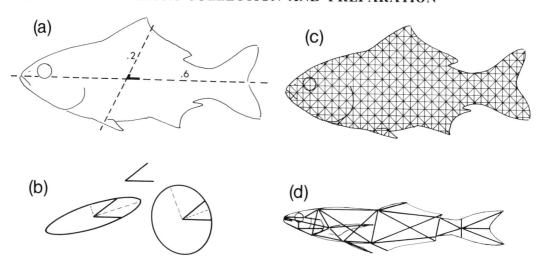

Figure 3.1.2 Loadings as descriptors of shape-change. (a) Loadings, corresponding to regression coefficients of length on size or shape differences, are presumed to apply uniformly at all points along their segments. Where two transects cross we have numerical information representing change in two directions. (b) Pairs of intersecting distance measures bear information about two directions only. We require three loadings to determine the parameters of even the simplest functions of orientation. Given two radii of an ellipse, for example, a third is needed to specify shape and size. (c) An ideal morphometric scheme would represent distances along closely spaced transects distributed evenly in direction. A gentle distortion of this design would be equally acceptable. (d) It is usually desirable that distance measures connect homologous landmarks. Transects should then be chosen to approximate even areal coverage and unbiased direction.

and to one particular landmark as well. Figure 3.1.4b shows a triangulation that is probably of least total length and perhaps of greatest generalized variance (Van Valen, 1974); Figures 3.1.4c and 3.1.4d show other possibilities of no particular formal distinction. All such schemes have the same number of distance-measures, $2n - 3$, and all bear exactly the same information: the total record of the point-configuration. Triangulated patterns are useful for descriptive measurements whether or not figures are meant to be reconstructed (Olson and Miller, 1958: fig. 55; Skalak et al., 1982).

We are not arguing that statistical analyses of forms proceed equally well using different arbitrary sets of triangulated distances as variables. The optimal measures of shape variation and shape difference which we ultimately derive usually involve distances oblique to those between homologous landmarks (Sec. 2.1). Further, it is not necessary that any measured distance be identical with "general size" as invoked in the literature. Size, a latent variable, is computed as a general factor; it is not measured (Sec. 4.2). The landmark configuration is most concisely and effectively archived without regard for the statistical uses to which derived measures might be put; the

3.1 Patterns of Distance Measurements

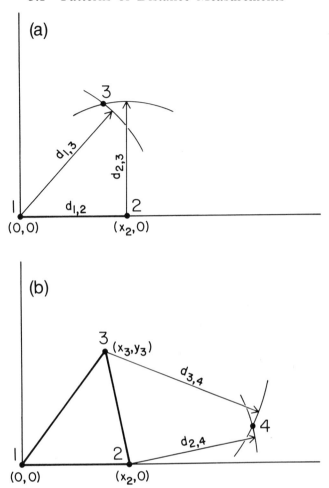

Figure 3.1.3 Mapping Cartesian coordinates of points by triangulation. (a) The first point is designated to be the origin of the coordinate system, and the line segment connecting the first two points to be the abscissa. The third point is positioned at the intersection of two circles, centered on the first two points, whose radii are its distances to these points. (b) The fourth point is then mapped in relation to the second and third. The procedure is continued until coordinates have been assigned to all points.

appropriate variables are best generated later in the research program. In the interests of serendipity, one might try to raise the probability of finding contrasts among the distances of the archive by designing a triangulation with short elements, a variety of orientations, and evenness of areal coverage. A good choice for our sample of seven landmarks might be the triangulation of Figure 3.1.4b.

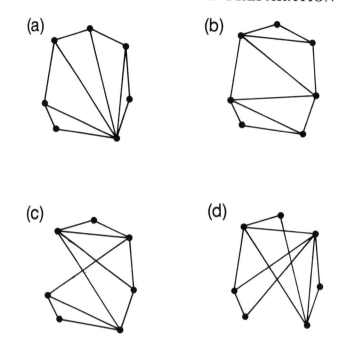

Figure 3.1.4 Examples of triangulation schemes for a set of seven landmarks.

3.1.3 Coordinate data

Another conventional archive is the collection of distances shown in Figure 3.1.5, distances measured from each of the landmarks to each of a pair of perpendicular lines. This is the familiar scheme of **Cartesian coordinates**; the figure exemplifies an arbitrary placement of axes. There are $2n$ distances here, three more than for the triangulation schemes. The extra three measures locate the axes with respect to the form (two for origin and one for orientation), information that is, in general, without biological meaning.†

Cartesian coordinates are often used directly as variables of a multivariate analysis. (Examples are cited in Bookstein, 1982a.) We consider this use of Cartesian coordinates, or any coordinates, to be inappropriate. The mathematical essence of a coordinate is the use of a number to name a line or curve. For instance, the x- and y-coordinates of a point on the digitizing tablet name the two lines, vertical and horizontal, at the intersection of which the point is located. Points on the same curve

---

† If one eliminates these three coordinates by setting the origin to one of the landmarks, and orienting one axis along the line from the origin through another landmark, then the construction is no longer symmetric because every landmark is now measured with respect to two in particular.

3.1 Patterns of Distance Measurements

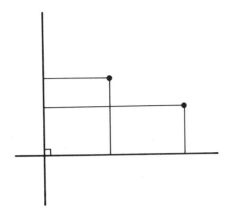

Figure 3.1.5 Cartesian coordinates as measured distances from points to each of a pair of perpendicular lines.

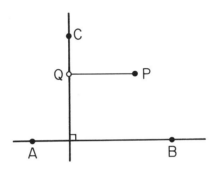

Figure 3.1.6 Cartesian coordinates are not homologous distances. Points A, B, C specify a Cartesian coordinate system with the x-axis along AB and the y-axis perpendicular to AB through C. The x-coordinate of point P is the distance PQ; but the point Q is not homologous from instance to instance of the form.

are *co-ordinated*; it is their assembly that is crucial, not their assigned axis values. For instance, the ordinary Cartesian pairs (x,y) which we use to label output from the digitizer are in most tablets names of actual physical wires. The numbers label these loci by conveniently ordering their intersections with some axis, but they need not.

A Cartesian coordinate is the distance from a point to a fixed line. The coordinate curves are these lines and their parallels. Unfortunately, of the two points between which distance is measured, only one is a landmark; the other is *biologically* arbitrary, being the foot of a perpendicular from landmark to coordinate axis. These coordinate axes are themselves arbitrary except possibly at the three points used to define them (Fig. 3.1.6). In other words, although Cartesian coordinates are indeed measured distances upon the form, they cannot be measured homologously.

Cartesian coordinates encode distances measured in two directions only, whereas sound morphometrics is grounded in a selection of distances in *all* directions (Sec. 2.1). For instance, the schemes of McCune (1981) or Fricke (1982), earlier proposed by Gregory (1928, 1933) and Taliev (1955), take landmark coordinates in an arbitrary system. The landmarks are thereby adequately archived; but nevertheless the coordinates separately, their ratios, etc., are of no particular value as variables.

Other sorts of coordinate systems, such as polar, elliptic, and the others mentioned in Appendix A.3, provide coordinate values having exactly the same flaw: they do not represent homologous measures, or a fair sample of directions, and therefore are not suitable as morphometric variables.

3.2 The Truss Network

Another pattern of distance measures using extra elements is the **box truss** (Fig. 3.2.1c). Use of this pattern allows us to: (1) systematically detect shape differences in oblique as well as horizontal and vertical directions, using a system of measures that ensures generally even coverage of the form within the landmarks; (2) archive the configuration of landmarks so that the form may be reconstructed (mapped) from the set of distances among landmarks; and (3) recognize and compensate for random measurement error.

3.2.1 Archiving the form with redundancy

Among the many possible patterns of distance measures that may be used to archive a configuration of landmarks, the simplest is triangulation (Sec. 3.1.2). When the measured distances among landmarks are chosen to form a series of contiguous triangles, the set of measurements may be used to map the coordinates of the landmarks (Figs. 3.2.1a and 3.1.3). The number of measures to be taken for a triangulation network increases as $2n$, twice the number of landmarks. However, an important problem attends the sequential mapping of coordinates by triangulation: small measurement errors are propagated as distortion throughout the form, so that the positions of the landmarks last located can be unreliable.

The accuracy of the mapping could be improved if redundant distance measurements for each landmark were used to average the effects of random measurement error. Such a system was described by Rohlf and Archie (1978) to map landmarks (in their case, the positions of trees in a forest) using repeated measurements of distances among them. In their system any three landmarks are chosen for a basal reference triangle. Distances are then measured from each additional landmark back to any three previous ones (Fig. 3.2.1b). From these data, initial estimates of landmark coordinates can be established by triangulation. However, because each landmark sits at the apex of one or more extra triangles, the

3.2 The Truss Network   53

(a) 2n-3 distances

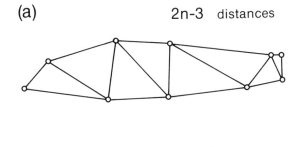

(b) 3(n-2) distances

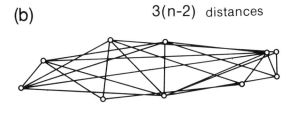

(c) 5n/2-4 distances

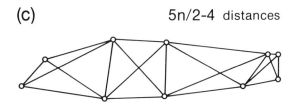

Figure 3.2.1 Possible patterns of distances which completely archive a configuration of ten coplanar landmarks. (a) A triangulation network. (b) A "globally redundant" design patterned after Rohlf and Archie (1978). (c) The box truss.

redundant measurements can be used to calculate an average set of coordinates by means of iterative least-squares. The technique can map landmark coordinates with a precision greater than that inherent in the original distance measures. The number of measurements needed increases as $3n$, which is still less than the $n(n-1)/2$ distances distributed among all possible pairs of landmarks.

The box truss (Fig. 3.2.1c) is a systematic pattern of measurements intermediate between these two. In this system, homologous landmarks on the boundary of the form are divided into two tiers and paired. (There may be a single landmark left over at one end or the other.) The distance measures are the **truss elements** connecting these landmarks into an overdeterminate truss network, a series of contiguous quadrilaterals each having both internal diagonals. Each quadrilateral shares one edge with the preceding quadrilateral and another with the succeeding one. In this way we add approximately one extra distance measure for each four present in a triangulated network. Paired landmarks at the ends of the truss network lie at the apex of three triangles; other paired landmarks lie at the apex of six triangles. This modest but systematic redundancy allows for checks on the consistency of the measurements (see the next section). The extra distances permit the positions of mapped landmarks to be averaged, limiting the accumulation of measurement error. If landmarks are spaced suitably on the outline, the pattern of measurements will approach an idealized square box truss (Fig. 3.2.2a), for which the expected confidence region for each landmark is smallest. The number of elements which must be measured for the truss without unpaired landmarks is $5n/2$, midway between the $2n$ characters of the triangulation and the $3n$ characters needed for Rohlf and Archie's "globally redundant" system. The pattern ensures balanced coverage across the form and can be applied to various projections which together account for three-dimensional aspects of shape (Fig. 3.2.2b,c).

3.2.2 Adjustment for measurement error: flattening the truss

For each quadrilateral of the truss network there are six distances (four edges and two diagonals) among four landmarks (Fig. 3.2.3a). For the landmarks to be precisely coplanar, these distances must satisfy a single equation—the determinant of the following matrix must be exactly zero (Salmon, 1914, v. 1:47):

$$\begin{vmatrix} 0 & 1 & 1 & 1 & 1 \\ 1 & 0 & d_{12}^2 & d_{13}^2 & d_{14}^2 \\ 1 & d_{12}^2 & 0 & d_{23}^2 & d_{24}^2 \\ 1 & d_{13}^2 & d_{23}^2 & 0 & d_{34}^2 \\ 1 & d_{14}^2 & d_{24}^2 & d_{34}^2 & 0 \end{vmatrix} = 288 \ V^2$$

Here d_{ij}^2 is the squared distance between landmarks i and j, and V is the volume of the tetrahedron they determine. Because these measurements will often be taken with calipers on pliable, hand-held specimens, it is unlikely that the six distances will exactly satisfy Salmon's criterion of planarity even though the landmarks would ideally lie in a single plane (e.g., the midsagittal). The distances will instead correspond to the six edges of a tetrahedron with some measurable volume (Fig. 3.2.3b). The mapping of landmarks then becomes a matter of flattening the tetrahedron by adjusting the six distances until the landmarks become coplanar.

Any five of the six distances exactly specify two coplanar triangles sharing an edge. We need not identify any particular one of the six distances (edge or diagonal) in a quadrilateral as the "extra" one. Given any five distances, the sixth must take one of two values, depending on whether the triangles are on the same side of the shared edge (Fig. 3.2.3c) or on different sides (Fig. 3.2.3d). For each of the six sets of five distances, the sixth distance will vary by measurement error (its own plus that due to error in the other five distances) from the exact value forced upon it by this planar construction. If

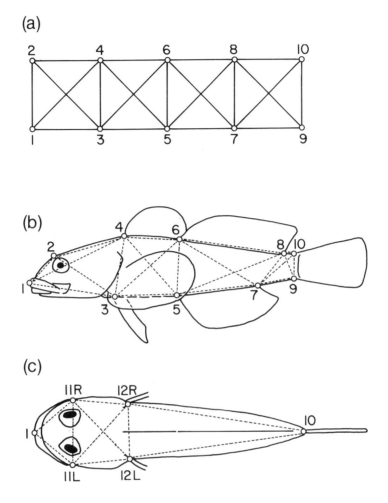

Figure 3.2.2 A truss for *Cottus*. (a) An idealized square truss for 10 landmarks. (b) A truss network of distance measures applied to 10 midsagittal landmarks of a sculpin (*Cottus*). (c) A single truss cell with two appended triangles, applied to the dorsal projection of the same organism.

the sixth length as measured is between the extremes of the two lengths indicated in Figures 3.2.3c–d, one can imagine this distance to be that of the real edge of a tetrahedron made up of the two triangles hinged along their common edge (Fig. 3.2.3b); the tetrahedron is the same for all choices of five edges out of six. The tetrahedron cannot be drawn if the length of the sixth edge is beyond either extreme, just as no triangle can be drawn with one edge-length greater than the sum, or less than the difference, of the other two; the "angle" at the hinge in these cases is imaginary, and the squared volume according to the preceding determinant is negative.

Cell by cell, whether or not the tetrahedron to be flattened is real or imaginary, we may compute the exactly planar configuration that minimizes the sum-of-squares

56   3 DATA COLLECTION AND PREPARATION

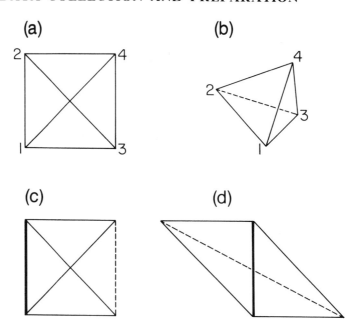

Figure 3.2.3 The geometry of flattening one truss cell. (a) A single truss cell consisting of four landmarks connected by six distance measures. (b) If the six distances do not satisfy Salmon's criterion of planarity, they will instead describe a three-dimensional tetrahedron. (c) For any set of five distances, the minimum permissible value of the sixth (dotted line) is the distance between the apices of two coplanar triangles lying on the same side of their shared edge (heavy line). (d) For the same set of five distances, the maximum permissible value of the sixth (dotted line) is the distance between the apices of two coplanar triangles lying on opposite sides of their shared edge (heavy line). If the length of the sixth distance is intermediate between these two extremes, the six distances describe a tetrahedron. If the sixth distance is beyond these extremes, the tetrahedron is imaginary.

$\Sigma_1^6 (d - \hat{d})^2$ between the measured distances and the edges and diagonals of the flat reconstruction. The fitted planar quadrilateral will approximately preserve $\Sigma_1^6 d$, the sum of the lengths in each cell, while balancing error-of-fit about all six distances. If one edge is grossly mismeasured, the fitted quadrilateral will show a pattern of compensating alterations in all the other distances. A combination of resistant fit (Siegel and Benson, 1982) and global affine fit (Tobler, 1977) might yield even sharper localizations of this error.

Each pair of landmarks at which two quadrilaterals abut will likely be adjusted to different distances in the two separate quadrilateral adjustments involving them. In the assembly of the final configuration this discrepancy, which is usually quite small, can be compromised in the following way. The first quadrilateral is laid down arbitrarily, with its lower left landmark positioned at coordinate (0,0). Each succeeding quadrilateral is then sequentially adjoined to its predecessor such that the midpoints of the shared edges are superimposed and the edge directions aligned. This superimposition continues to approximately preserve the sum of the truss distances, and adds least error to the distances computed for those edges separately (at slight cost in net accuracy for the horizontal and diagonal segments).

The reconstruction of an entire truss in this way closely approximates a global best-fit that is feasible but much more expensive to compute. Propagation of errors from cell to cell (e.g., Fig. 3.2.4d) is much less extensive than for triangulation without redundancy. The cross-measures we take tend to be not only shortest but also most nearly at 45° to the anteroposterior and dorsoventral measures, and therefore bear the most additional information. The longer distances that might be used to augment a globally redundant best-fit, such as diagonals 1–6, 2–5, or 3–10 (Fig. 3.2.2c), neither add much robustness to the reconstruction of forms as elongated as most fishes, nor contribute much additional information to subsequent morphometric analysis. Their loadings are extended averages of coefficients already quantified locally.

3.2.3 Examples

Figure 3.2.4 shows five specimens of the freshwater sculpin *Cottus cognatus* for which the landmarks have been mapped by this reconstruction. All except the fourth are reasonable approximations to the form of a fish. The apparent distortion of this individual is the result of a gross measurement error which was immediately obvious once the figure was drawn. Many such errors, especially those resulting from misreadings of calipers, ocular micrometers and other measuring devices, may go undetected in morphometric studies even when they result in outliers on scatter plots or histograms.

Substantial measurement errors will be exposed when the form is drawn, while more subtle errors may be revealed by measures of the mutual inconsistency of the data. A useful measure of such **strain**, adjusted for the varying sizes of the cells, is the relative discrepancy between measured and reconstructed distances, $(d - \hat{d})/d$. (Note that this is not quite the quantity that is minimized in the flattening.) This statistic, computed for each distance measure, may be aggregated to describe the strain for each quadrilateral, each specimen, or for a sample of forms. One satisfactory net measure of strain is the root sum-of-squares of these, $\sqrt{\Sigma_1^{21}[(d - \hat{d})/d]^2}$, for all the distances of a truss. A second useful measure of strain, at the level of the cell rather than the edge, might be the volume V of the tetrahedron before it is flattened, scaled by the 1.5 power of the flattened area. Many others are possible.

The distance deviations (Table 3.2.1) for the trusses of Figure 3.2.4 indicate that the distortion of the fourth specimen may be the result of a measurement error for the lower horizontal element (between landmarks 3 and 5) of the second truss cell. In fact, this distance was undermeasured by approximately 9%; the lengths of the other distances in the cell were adjusted to compensate for this. The values of strain also reveal that for all five specimens the positions of landmarks 9 and 10 are somewhat

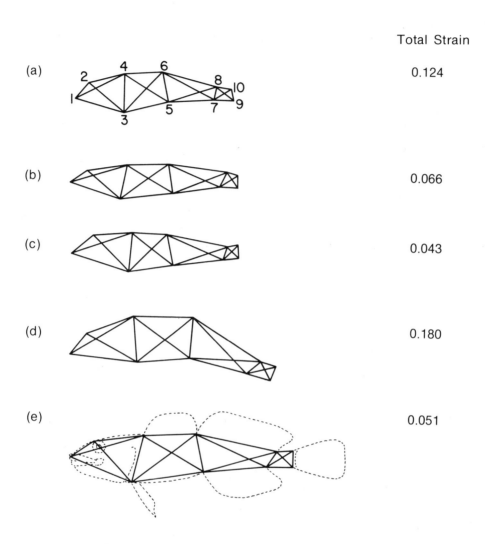

Figure 3.2.4 Mapped landmark configurations for five specimens of the sculpin *Cottus cognatus*. Landmarks are those indicated in Figure 3.2.2c. The body form of the fish in relation to these landmarks is indicated on the fifth figure. Total strain is a measure of the mutual lack of fit of the original distance measurements; it is the square root of the sum of the squared relative deviations of the distances in the mapped form from those originally measured.

Table 3.2.1 Strain statistics for the five trusses of Figure 3.2.4. The value for each distance measure (identified by the landmarks it connects) is the difference between the original measure and the corresponding distance on the final mapped configuration, scaled by dividing by the original measure. Net strain for each cell is the root sum-of-squares (RSS) of the six strain values. Net strain per form is the RSS of 16 or 21 strain values.

Specimen	Cell	Distance Measures						RSS
		1-2 3-4 5-6 7-8	1-3 3-5 5-7 7-9	2-3 4-5 6-7 8-9	1-4 3-6 5-8 7-10	2-4 4-6 6-8 8-10	3-4 5-6 7-8 9-10	
1	1	−.007	−.001	.004	.005	−.004	.005	.012
	2	.005	.012	−.018	−.018	.018	.010	.036
	3	.010	.005	−.002	−.005	.004	−.016	.020
	1-3							.041
	4	−.016	−.034	.048	.048	−.067	−.059	.118
	1-4							.124
2	1	.003	.001	−.001	−.001	.002	.001	.004
	2	.001	.002	−.002	−.002	.003	.004	.006
	3	.004	.019	−.012	−.017	.008	.005	.030
	1-3							.031
	4	.005	−.007	.025	.025	−.038	−.026	.058
	1-4							.066
3	1	.008	.002	−.003	−.005	.008	.004	.013
	2	.004	.006	−.010	−.009	.011	.005	.019
	3	.005	−.002	.002	.002	−.001	−.010	.011
	1-3							.025
	4	−.010	−.008	.014	.015	−.024	−.013	.036
	1-4							.043
4	1	.011	.001	−.011	−.010	.004	−.033	.038
	2	−.033	−.090	.078	.078	−.062	−.042	.164
	3	−.042	.007	−.009	−.005	−.004	−.010	.045
	1-3							.166
	4	−.010	−.013	.027	.029	−.048	−.027	.072
	1-4							.180
5	1	.015	.003	−.005	−.008	.010	−.002	.021
	2	−.002	−.004	.008	.008	−.010	−.004	.016
	3	−.004	−.000	.000	.000	−.001	−.010	.010
	1-3							.027
	4	−.010	−.008	.016	.018	−.032	−.015	.044
	1-4							.051

uncertain, resulting in inconsistencies among the measures involving them. In such circumstances the affected truss cell might be excluded from subsequent analyses.

As a second example of the use of such statistics, we have plotted against a composite measure of body size the total strain per individual for data from 43 specimens of another sculpin, *Cottus klamathensis* (Fig. 3.2.5). Although only a slight relationship of imprecision of measurement to body size is evident,† there are obviously several individuals for which the data fit poorly. Three of the four data points of highest strain corresponded to specimens for which either caliper settings were misread or keying errors were made. Values of strain for each distance measure within the forms again indicated the particular measures that were in error.

To test the degree to which the least-squares mapping procedure corrects for imprecision of measurement, we took data on a single specimen of *Cottus cognatus* to the nearest .05 mm, the usual resolution of measurement, and mapped them to reconstruct the form. The data were then remapped after being rounded to the nearest millimeter (Fig. 3.2.6). The strain estimate for the less precise data is twice that for the data as originally measured, 0.121 versus 0.073, but the resulting forms are virtually identical. We then repeated the procedure for nine more specimens. Figure 3.2.7 is a scatter of the Cartesian coordinates of the ten individuals mapped at the lower precision (mean strain, 0.103) against the same coordinates mapped at the higher precision (mean strain, 0.058). If the mapped forms were identical over the refinement, the points would lie exactly along the 45° lines of equality. In fact, even though the strain is consistently greater for the less precise data, the points lie very close to this line; the slight distortion created by rounding off the data is corrected by the least-squares algorithm.

The system of measurement we have described is based on two premises: first, that collections of anatomical points and the distance measures among them must be homologous from form to form, because only the biological homology of two configurations makes meaningful their scientific description and comparison; and second, that an adequate collection of measured distances should at least permit the reconstruction of the configuration of landmarks it purports to measure, because information is otherwise lost. The truss network is a geometric protocol that fulfills these requirements and offers several additional advantages. Its proximate advantages may be summarized as follows; other properties will be described in Sections 4.4 and 4.6.2.

1. Use of the truss network as a character set enforces systematic coverage across the form; in contrast, traditional character sets often provide highly uneven coverage.

2. The truss exhaustively and redundantly archives the form; hence, the original configuration of landmarks may be reconstructed by relaxing the data so as to be coplanar and mapping the Cartesian coordinates of the landmarks.

† In the absence of measurement imprecision, strain is expressing nonplanarity of the original configuration of landmarks, and so would not be expected to correlate with specimen size.

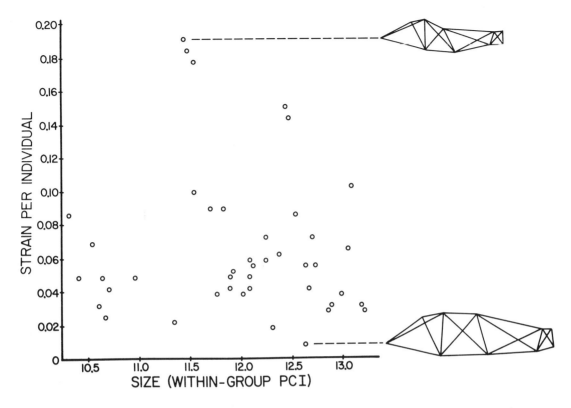

Figure 3.2.5 Scatter plot of strain per individual against a composite measure of body size, for 43 specimens of *Cottus klamathensis*. Body size is the score on the first within-group principal component of the covariance matrix of log-transformed distances. Reconstructed forms are shown for the two specimens having the best and worst fit.

3. The least-squares mapping procedure is robust against moderate imprecision of measurement. The degree of measurement error in the data may be assessed graphically and by various indices characterizing the mutual lack of fit of the data.

3.3 Digitizing

In the truss method, a configuration of landmarks is stored by way of certain distances among them. An alternative, already mentioned in Section 3.1, is the archiving of landmark positions one by one in terms of their coordinates. Except in the very simplest applications to sets of three or four landmarks, one should digitize at a **digitizing tablet**, a device that reads Cartesian coordinates at the location of a **cursor** (electric pen, cross-hairs, or the like) and transmits them to a computer memory. Three-dimensional digitizing of exterior points may be effected by cursor, camera(s),

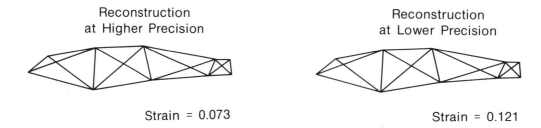

Figure 3.2.6 Reconstructed truss networks for a single specimen of *Cottus cognatus*. Measurements were originally taken to the nearest 0.05 mm (higher precision), and then rounded to the nearest 1 mm (lower precision).

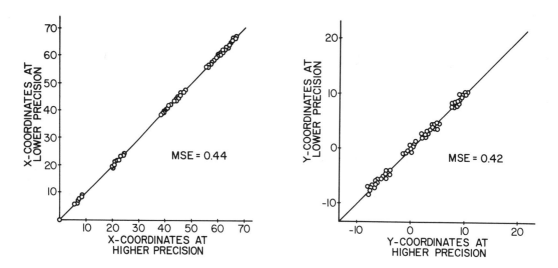

Figure 3.2.7 Scatter plots of the Cartesian coordinates of reconstructed truss networks for 10 specimens of *Cottus cognatus*. In all the reconstructions, landmark 1 was set to the origin and landmark 9 placed along the *x*-axis. Measurements were originally taken to the nearest 0.05 mm (higher precision), and then rounded to the nearest 1 mm (lower precision). If the mapped forms were identical over the change in precision, the points would lie exactly along the 45° lines of equality. Axis scales differ for the two plots. The mean-square errors (MSE) of the regressions are approximately equal.

or acoustics. Most computer-graphics facilities have programs that manage data streams from the digitizer and that allow specification of Cartesian coordinate reference axes and inclusion of specimen names or scale information. Some devices allow a choice between "point" and "stream" mode. In "point" mode, the researcher consciously chooses each location to be transmitted. Data collected in stream mode should subsequently be filtered according to the rules suggested in the next section.

3.3 Digitizing

3.3.1 Forms without landmarks

The **boundary curves** of forms without landmarks, such as the opercular outlines of Section 4.1, are digitized as a string of arbitrarily many points and interpreted as the polygon having those points for vertices.

One may begin digitizing these forms anywhere on the outline—when the polygon is closed upon itself the identity of the starting point is moot. In practice the starting point is marked with a pencil on the raw image. The researcher positions the crosshairs of the cursor at that point and proceeds in a consistent direction around the outline, selecting points at the center of the line being traced and at a spacing approximately proportional to degree of curvature (Fig. 3.3.1). In this way the sequence of short segments connecting each point to its two neighbors is made a fair representation of the outline. We think this protocol produces better records than do systematic samples of the boundary curve taken evenly either in arc-length (e.g., every 2.5 mm) or in angle (e.g., every 3° with respect to some center). In this style of sampling, each point tends to bear roughly the same amount of geometric information; in systematic sampling, on the other hand, certain areas are undersampled unless all areas are oversampled.

During the digitizing it is helpful to glance a few centimeters ahead of the cursor and ensure that among the points transmitted are corners and tips of processes, both internal and external, and points of inflection: that is, local maxima and minima and zeroes of curvature. Should one not wish serrations or other small-scale irregularities of the boundary to affect results, the transcription may be smoothed by selecting points roughly halfway from bottom to top of the serrations. Likewise, there is no harm in skipping rather long sections of the outline if the resulting representation as a straight-line segment is adequate for the purposes of the analysis. End the digitizing when the cursor returns to the first point digitized.

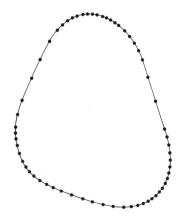

Figure 3.3.1 Points to be digitized. Spacing of the actual points selected for digitizing should be approximately inverse to the sharpness of boundary curvature.

Any hand-digitized record of a continuous curve will include some points that deviate from the intended smooth curve by rather more than the width of the line. These points interfere with analyses such as the medial-axis algorithm that explicitly assume smooth trends in curvature of the boundary. We have developed interactive editing programs that permit us to "smooth" the digitization at the terminal when the programs seem to be misconstruing the original record for this reason.

3.3.2 Forms with landmarks

We archive the boundary curve of a form *with* landmarks as a curvilinear polygon of arcs between successive **boundary landmarks**. Far fewer actual points of the outline are stored than in the landmark-free representation. The purpose of landmarks on an outline is to coordinate outlines between forms, to calibrate a **boundary homology function** relating them. This requires a smoothness to the boundary which can be guaranteed only by long arcs rather than short jagged lines. We are really *simulating* the outline by a particular sequence of lines and curves, a sequence identical from specimen to specimen of a data set; only the parameters of the curves change. Hansell et al. (1980) refer to the resulting polygon as the **analytic boundary** in contrast to the **empirical boundary** as it would be drawn or digitized landmark-free.

The language we use for description of the arc scheme (see Sec. 4.5) offers four alternative descriptions for the boundary arc between a pair of landmarks (Fig. 3.3.2). Most involve **helping points**, which are auxiliary points digitized along with the landmarks to aid in the reconstruction of the curving biological outline. A helping point has only one meaningful coordinate, position normal to the line through its neighbors; once it is used to determine the location of a boundary arc, knowledge of its position is discarded, so that it bears no information about computed-homology along the arc. The arc may be

1. a straight segment,
2. a circular arc constrained to pass exactly through a helping point,
3. a parabolic arc defined by its tangents at both bounding landmarks, or passing through two helping points, or
4. a general **conic arc**—ellipse, parabola, or hyperbola—constrained by three helping points or by its tangents at both bounding landmarks and one further helping point.

The conic arcs drawn between landmarks are consistently convex or concave to the straight segment across the endpoints. (Their curvature never changes sign.) This makes these curves inconvenient for the representation of empirical arcs with points of inflection (at which, by definition, the curvature passes through zero on its way from positive to negative). As the tangent line may pass from one side of the boundary to the other only when we change from one conic arc to another, we expressly digitize the

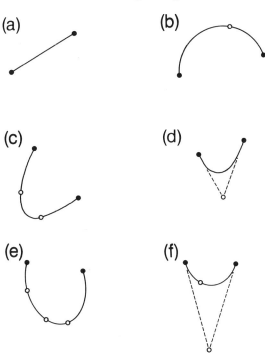

Figure 3.3.2 Varieties of boundary arc between landmarks. (a) Straight segment. (b) Circle, specified by one helping point on the curve. (c) Parabola, specified by two helping points. (d) Parabola, specified by one helping point at which the tangents through the landmarks intersect. (e) Ellipse (or hyperbola), specified by three helping points on the arc. (f) Ellipse (or hyperbola), specified by one helping point on the arc and one point at which the tangents through the landmarks intersect.

important inflections as helping points separating arcs of opposite curvature. Along with all the other helping points, these are ignored in any computations of homology along the boundary.

The purpose of digitizing in the landmark-free style is to fairly sample both the points and the tangents of the outline data. The purpose of the second, landmark-based style is quite different: it is to characterize the interior of the form, the domain of the homology function. Whereas the careful location of landmarks is a more or less straightforward task, the location of the helping points is subjective. Some of these points, such as the intersections of the tangents which may have been assigned in versions (3) or (4) above, are not upon the form at all. Even those through which the arcs are constrained to pass need to be selected carefully so that the curve they help define will approximately reproduce important features of the empirical outline. The judgment of the quality of this approximation is a matter for experience, not for statistics.

3.3.3 Example: Atherinids

As an example of the rationale underlying a digitizing scheme we shall explain the stylization of the atherinids analyzed in Section 4.6.1. Figure 3.3.3 presents a typical drawing of the sort we wish to digitize. We may ask first whether we wish the "form" to include the fins. For analysis of change as deformation we believe it should not, because the fins flex and erode independently of other contours, and because their "interiors" do not pass homogeneously into the body of the fish. (Allometries involving their lengths may be entered later in the analysis.) For the ten points shown, there is good reason to believe that apparent homologies are real—that their apparent displacement with growth is true identity of tissue, not redefinition or remodeling. We select these ten points as the landmarks of the outline.

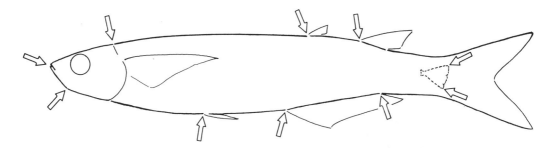

Figure 3.3.3 Schematic of the atherinid data base. (This is used in the example of Section 4.6.1.) Landmarks, clockwise from left: tip of snout; posterior border of the skull; origin of the spinous dorsal fin; origin of the second dorsal fin; top, bottom of the hypural plate; posterior base of the anal fin; origin of the anal fin; insertion of the pelvic fin; articulo-quadrate joint.

Inspection of this and other specimens in the collection suggests that we can reasonably replace three arcs of the outline by straight segments: origin of spinous dorsal fin to origin of second dorsal fin, length of base of the anal fin, and snout to articulo-quadrate joint. (See Figure 3.3.4.) Three other arcs curve in such a way that they can be represented as segments of circles: from the snout to the posterior border of the skull; from there to the origin of the spinous dorsal fin; and the posterior margin of the hypural plate. To define each of these arcs we digitize an additional helping point c somewhere along its span, usually near the middle, chosen carefully so that the circle lies near all of the boundary points it is supposed to summarize. We sometimes experiment with the location of the helping point until we arrive at a subjectively satisfactory arc.

Two other arcs, bounding the caudal peduncle ventrally and dorsally, are definitely not circular. They curve more sharply posteriorly than laterally, and can be represented by segments of parabolas. These are the algebraically simplest curves bearing appropriate tangents at these two bounding landmarks. To this end we

3.3 Digitizing 67

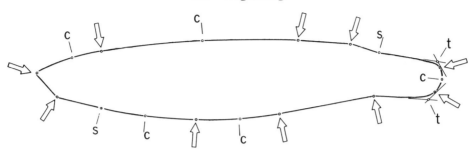

Figure 3.3.4 Digitizing scheme for Figure 3.3.3. Ten landmarks (arrows); five helping points (*c*) for circular arcs; two intersections (*t*) for tangents of parabolic arcs; two splines (*s*) between circular and straight subarcs between landmarks.

construct (by hand) the two points *t* which are the intersections of the pairs of tangents at the endpoints of the arcs. These points provide the additional information necessary to determine the parabolas passing through the landmarks and having the specified lines as their tangents at the landmarks.

The two remaining arcs, from the origin of the second dorsal fin to the top of the hypural plate and from the insertion of the pelvic fin to the articulo-quadrate joint, require composite definitions. Each is the assembly of one straight segment with a gently curving arc. The two loci *s* of transition are named "posterior base of the second dorsal fin" and "ventral symphysis of the cleithra"; but we do not consider them to be homologous from form to form—they are **pseudolandmarks**, helping points of another sort. In each of these arc-pairs the gently curving part is modeled as circular; therefore we are required to digitize one additional helping point per span.

There results the digitizing scheme of Figure 3.3.4, involving a total of 19 carefully selected points: ten anatomical landmarks, five helping points for circular arcs, two intersection points of tangents to parabolic arcs, and two pseudolandmarks serving as separators between straight and circular segments. The analytic boundary specified by this particular set of 19 points may be compared with the original outline form. Naturally the fit is close—it was designed to be. To the extent that other specimens of the data set are similarly shaped, these 19 points capture all the information we need for comparisons among the outlines.

Comparing atherinids to fish of rather different form would require a different assemblage of arcs, or perhaps could not proceed beyond the representation by landmarks and straight segments. The context of biological homology demands one-to-one correspondence of curvatures, as well as landmarks, from form to form. Of course, neither the point-coordinates nor the coefficients in the equations of the arcs are tractable parameters for the comparison. (Recall Section 3.1.3.)

4

Analytic Methods

The preceding sections set forth protocols for gathering morphometric data archives of four different types: curving forms without landmarks, lists of measured distances between landmarks, digitized locations of landmarks, and curvilinear outlines augmenting landmarks by conic arcs connecting them. In this chapter we work outward from the discipline of these archives toward demonstrations of new or modified analytic methods.

For the first sort of data set, the curving form without landmarks, Section 4.1 describes Blum's method of medial axes, a technique that characterizes forms as the aggregations of empirically demarcated parts and regions. Lists of interlandmark distances, the second form of data base, omit just this information about curving form; nevertheless, their analysis should express the original geometric ordering underlying the measurements. In our view, Sewall Wright's general method of path analysis is the only multivariate approach that can take into account that geometric origin. We explain the principles of this method in Section 4.2, and present in Section 4.3 a specific application involving the simultaneous consideration of factors for size and for shape-change: discrimination among groups that vary in size.

The remainder of the chapter deals with the third and fourth forms of data, based upon the explicitly geometrical information of landmark locations. Section 4.4 shows how the truss data structure we have recommended, together with Wright's model of size allometry, allows the convenient averaging of geometric forms spanning an extensive range of biological sizes. For such averaged forms, as for "typical" forms selected in other ways, the method of biorthogonal grids makes full use of any geometric information available—landmark locations, if that is all there is, but also information on the curving of form between landmarks. Section 4.5 explains how the notion of principal directions of change (Sec. 2.1) is extended into a complete coordinate system, the biorthogonal grid, custom-fitted to any observed smooth shape comparison. Section 4.6 demonstrates the interaction of biorthogonal analysis with path-analytic factor analysis in the consideration of form-change between species simultaneously with that within species. Finally, in Section 4.7 we revert to the larger systematic context, introducing the analysis of pairwise comparisons, both ontogenetic and phylogenetic, for forms of several taxa archived by landmark locations.

4 ANALYTIC METHODS

4.1 The Medial Axis

4.1.1 Principles

The **medial axis**, or **symmetric axis** or **skeleton**, of a curving form is a sort of stick figure for the extended shape, a precisely defined middle curve. The best introduction to this notion is through two major essays by Harry Blum (Blum, 1973; Blum and Nagel, 1978). Blum, a visual psychologist, advanced this formalism in the early 1960's as a "biological geometry" for the study of "amorphous forms." Ironically, it was noticed only in the then-new field of military computer vision, where attempts were made to apply it to classic problems of pattern-recognition—jet fighters, tanks, etc. Apparently unsuited for that purpose, it was discarded before 1970. Only recently has it been resurrected to enter the biometric literature for which it was originally intended. For an up-to-date exposition of the application to morphometric texture analysis, see Serra (1982). Our own intuition of this concept, appropriate to higher biometric applications, is taken from Bookstein (1981a).

The middle of any outline form, as we operationalize it, should be equidistant from its sides. The distance measure invoked here is the distance of a point from a curve. This distance is the radius of a circle around this point, the circle that just touches the curve (Fig. 4.1.1a). A point is in the middle of a whole curving form, in this sense, only if it is at the center of a circle that touches the boundary in two separate places. The medial axis of an outline is simply the collection of all the points that are "in the middle" in this way—all centers of circles that touch the outline at two distinct points (Fig. 4.1.1b). The medial **width** of the form at any medial point is the radius of the circle centered there.

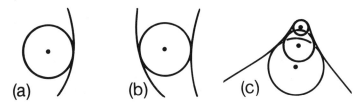

Figure 4.1.1 The medial axis. (a) The distance of a point from a curve is the radius of a circle touching the curve at one point. (b) A middle point of a form is the center of a circle touching the boundary curve at two or more points. (c) The endpoint of a string of middle points (the medial axis) occurs at a local maximum of curvature.

Medial axes start (or stop) at points where the two associated boundary arcs abut. Such **end points** correspond to points of the outline at which the curvature is a local maximum: tips or corners of the form, as in Figure 4.1.1c. Axes branch wherever a third boundary arc intrudes into a sequence of circles touching a pair of sides (Fig. 4.1.2). From the **triple point** emerge three separate segments of the medial axis, corresponding to the three ways of choosing two out of the three boundary elements

4.1 The Medial Axis

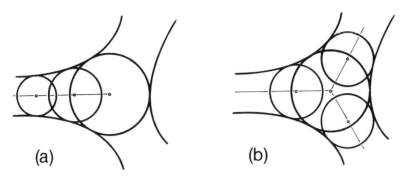

Figure 4.1.2 Triple points in a medial axis. A branching point of the medial axis corresponds to the center of a circle that touches the boundary curve at three points, such as when a third boundary arc "intrudes" on a series of circles (a). From this triple point emerge three segments of the medial axis, corresponding to the three arms of the form (b).

now involved. It makes no difference which side was the "intruder": the three branches meeting at a triple point have no formal sequence.

Medial axes for familiar forms have an appeal corresponding to aspects of lower-level visual perception (Fig. 4.1.3, redrawn from Blum, 1973). It is possible to make a form symmetric by straightening its symmetric axis. Blum proves that this transformation leaves both area and perimeter unchanged. Hence it may provide a convenient normalization for specimens preserved or observed in arbitrary states of flexure.

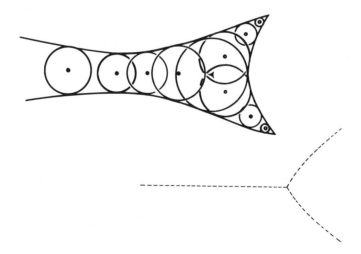

Figure 4.1.3 A simple medial axis. The medial axis often corresponds to a "stick-figure" of a recognizable object. After Blum, 1973.

The medial-axis computation accepts only whole curving shapes as input. Landmarks, even if available, are not used. For digital computing, the data take the form of an ordered series of ordinary points sampled all around the perimeter. The choice of starting point is irrelevant. Although the curve is thereby replaced by a polygon of straight chords, information about curvature is still present, implicit in sequences of points a few at a time. In the following example, Section 4.1.2, the polygonal outline data were transformed into medial axes by the line-skeleton algorithm of Bookstein (1979). It is also possible to compute these configurations by hand. Using a template of concentric circles (such as transparent polar coordinate graph paper) having a pinhole at its center, one can explicitly mark the centers of circles that appear to be tangent to the boundary at two separated points, and afterwards link up these dots into segments of the medial graph. Triple points will be suggested by intersections of medial arcs; they should then be located with care by the same template. It is helpful to practice this manual technique on structures whose symmetric axis geometry is obvious, such as the outline of a hand with fingers spread.

Corresponding to the assembly of any medial axis out of arcs is a decomposition of the form into regions (Fig. 4.1.3). Blum's essays, cited above, classify the possible shapes of these regions from an algebraic point of view. Blum and Nagel (1978) indicate how to suppress small branches of the medial axis corresponding to short arcs or small distortions of the boundary. Mathematics aside, pieces of the resulting dissection usually correspond to well-known anatomical parts. For instance, in a human mandible projected laterally (Fig. 4.1.4), the arcs of the medial axis correspond to the parts conventionally named condyle, coronoid process, ramus, gonial region, corpus, and symphysis. The medial axis delineates these parts purely geometrically, without any information about biological homology, and generates its own "landmarks" and orientations, the triple points and axis directions, by relating "opposite sides" of the form in twos and threes. Notice, for instance, how the endpoints of the medial arcs extend into the culs-de-sac of the form, twisting to align with the boundary points of tightest curvature.

From this analysis can be generated many new measures independent of conventional landmark schemes. For example, we might extract the widths (radii) of the form at the triple points, and also the lengths of the medial arcs connecting them. Whenever the topology of the medial axis is sufficiently reliable, the resulting measured lengths constitute a data matrix which can be transformed to the logarithm and used in allometric modeling or shape discriminations. Or we might sample the angles and curvatures throughout the system of medial arcs and branches (Webber and Blum, 1979). It is also possible to eschew all separate measurements and still exploit the medial axis for the direct visualization of change, by superposition of successive forms upon corresponding arcs or points of the axis (Bookstein, 1981a): see, for instance, Figure 4.1.5.

Even though there is no prior information about homology in the medial axis, the correspondence it suggests between geometrically analogous parts of extended forms may be a reasonable guess at a homology function. Beginning from mere outline data,

4.1 The Medial Axis

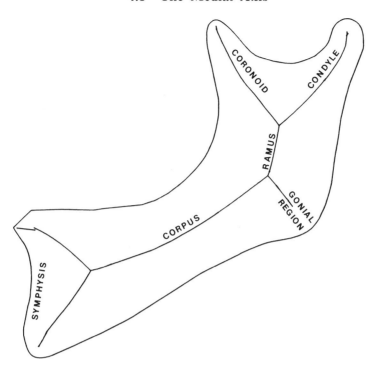

Figure 4.1.4 Correspondence of anatomical regions and medial axes. Projection of a human mandible showing labelled regions corresponding to parts of the medial axis.

the computation passes to interior parameters that seem appropriate for the description and comparison of shapes in many different contexts. The distribution of arcs out of a triple point in bone, for instance, may express an equilibrium of various biomechanical forces each associated with one of the bony processes, and so should manifest certain invariants over ontogeny (Webber and Blum, 1979). Blum (1973) argues this naturalness most persuasively. Because the circles and their radii capture all the information to be had in the absence of a homology function, all systematic aspects of form-change must be expressed somehow in the parameters of the medial axis. With experience, the investigator can select useful parameters after a study of a few examples across the range of natural variation. The technique is informal, subtle, and richly suggestive. We consider it the best available technique for studying variation in extended shapes lacking any obvious point homology function.

4.1.2 Example: Opercles

In the following example, we computed medial axes of 11 right opercles (Fig. 4.1.6) from two species of bass of the family Centrarchidae: *Micropterus dolomieui*, the smallmouth bass, and *M. salmoides*, the largemouth bass. In addition, two

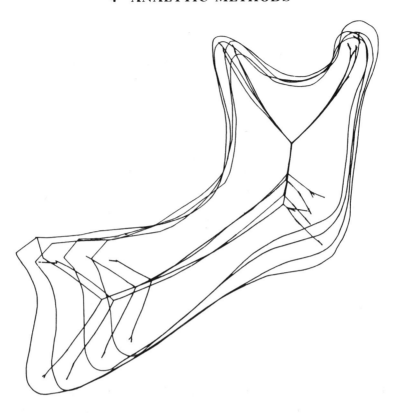

Figure 4.1.5 Change in a medial axis with growth. Superimposition of a growth series of projected outlines of human mandibles demonstrating changes in the medial axes over time. (From Bookstein, 1981a: fig. 4a.)

ontogenetic growth series were constructed by tracing annuli (Fig. 4.1.7) in the largest individual specimen of each species. Standard lengths of individuals of *dolomieui* ranged from 84 to 256 mm, of *salmoides* from 84 to 235 mm.

We can compare the medial axes in two ways: by superposition of axes of pairs of specimens (Fig. 4.1.8) and by measurement (or computation; see Appendix A.5) of angles among segments of the axes. The points of reference for either procedure will be the triple points and the extension of axis segments until they touch the boundary; in particular, angles must be measured at triple points. But individual triple points are not to be regarded as homologous from form to form whenever they derive from different regions of the outline. Therefore the angles at these triple points may require modification before their use as variables.

These opercles could be described by four sides nearly tangent to the same circle (Fig. 4.1.9). Their medial axes therefore involve pairs or trios of triple points situated very near to one another and approximating quadruple or even quintuple points. In order to compensate for this clustering of triple points, we measured angles at the centroid of the pair or trio by simple averaging of the coordinates (Fig. 4.1.10).

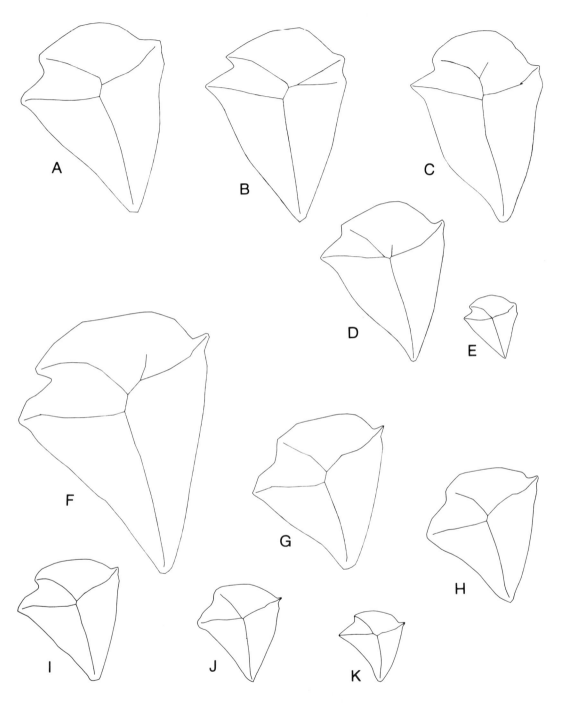

Figure 4.1.6 Size series of opercular outlines with medial axes. Digitized outlines and computed medial axes of 11 right opercles. Opercles A through E represent a size series of individuals identified as *Micropterus dolomieui*. F through K represent a size series of individuals identified as *M. salmoides*.

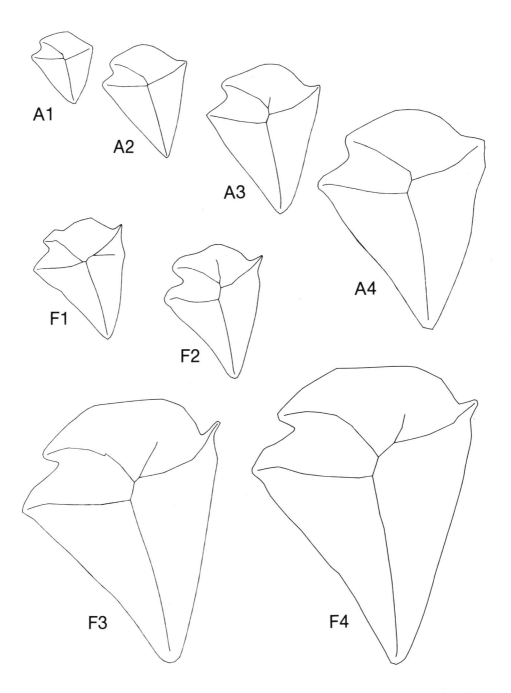

Figure 4.1.7 Growth series of opercular outlines with medial axes. Digitized outlines and computed medial axes of two growth series traced from the annuli of the largest opercles of each species: A, *M. dolomieui* and F, *M. salmoides*.

4.1 The Medial Axis

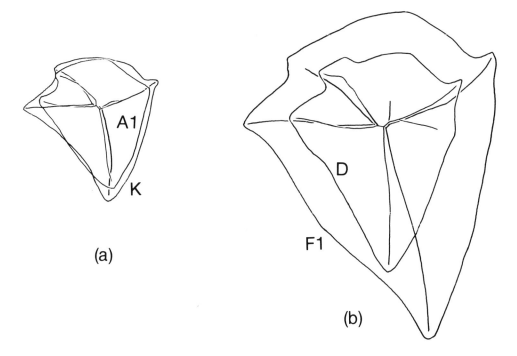

Figure 4.1.8 Superimposition of opercular outlines. (a) Superimposition of outlines A1, K with medial axes oriented by endpoints of the axes; (b) superimposition of outlines D, F1 with medial axes oriented by triple points.

The two species of bass are unambiguously discriminated by the angle at the centroid, as shown in Figure 4.1.10 and Table 4.1.1. In *M. salmoides*, the angle varies from 90° to 97° (mean 93.9°, s.d. 2.3°); in *M. dolomieui*, from 98.5° to 110° (mean 104.2°, s.d. 4.3°). Figure 4.1.8 illustrates this angular discrimination: comparably sized specimens of the two species are superimposed appropriately at the constructed point. There was one apparent misclassification. An opercle of *M. "dolomieui"* (specimen K) bore a measurement of 94.5°, within the range of specimens of *M. salmoides*. Examining other skeletal elements, we discovered that this individual was in fact a misidentified individual of *salmoides*. Thus the medial axis method allowed identification of individuals of closely related species, based only on the angle between two segments of the axis.

Trends within species are not entirely clear. Neither of the size series show obvious rotation of axes or consistent change in the measured angle. However, in the ontogenetic series of both species, the angle diminishes with age: from 97° to 94° in *salmoides* and from 108.5° to 98.5° in *dolomieui* (Table 4.1.1).

Figure 4.1.9 Close approximation of triple points in medial axes. Outline of opercle F3 with two circles centered on the pair of triple points. The near congruence of the two circles approximates one circle touching the outline at four points.

4.2 Path Analysis in Multivariate Morphometrics

In this section we introduce **path analysis**, Sewall Wright's (1932, 1954, 1968) powerful technique for analysis of covariance structures, and use it to illuminate or replace some conventional procedures.

You are accustomed to explanations of multivariate techniques as extractions of the linear combinations that optimize some quantity (e.g., a multiple correlation or an F-ratio). Multiple regression "explains" the most variance of the dependent variable by a linear combination of the predictors; principal components analysis explains the most variance of all the variables considered together by a sequence of linear combinations; multiple discriminant analysis "best separates" groups, again by a sequence of linear combinations. While conventional multivariate techniques explain or optimize variance, path analysis explains covariances and interprets certain equations, which may not correspond to the quantities being optimized in the conventional techniques.

4.2 Path Analysis in Multivariate Morphometrics

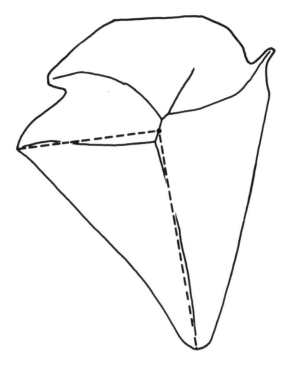

Figure 4.1.10 Angular measurement on a medial axis. Outline of opercle F3 showing the constructed point from which angles were measured.

4.2.1 Regression models and the consistency criterion

To understand this distinction in practice, let us review ordinary multiple regression. In the usual presentation it is described as a matter of "explaining variance." We are to compute the linear combination $a_1X_1 + \ldots + a_nX_n$ of predictors $X_1, \ldots, X_n$ that correlates best with a dependent variable Y. The a's, in other words, are the coefficients for which the variance of the linear combination $Y - a_1X_1 - \ldots - a_nX_n$ is minimal. This criterion leads to a system of linear equations, the so-called **normal equations**, of which the a's are the solutions.

But we can arrive at these same equations from a wholly different point of view (Bookstein, 1980b, 1982b). Let there be three predictors X_1, X_2, X_3, each of variance 1, and a dependent variable Y, also of variance 1. Then we can diagram our model for the regression of Y upon the X's as in Figure 4.2.1. There → means "contributes to" (in a causal sense, as by way of controlled experimental manipulation) and ↔ means "is correlated with" (in a non-causal sense). Such sketches are called **path diagrams**. We assume that all covariances among the X's and between them and Y are known, i.e., that we have no concern for sampling variation. That is, we know all the simple regression coefficients (correlations) r_{Yi} of Y on (with) each X_i, and also the

Table 4.1.1 Medial axis angles of bass opercles.

Species	Designation	Measured Angle	Dorsoventral Length (mm)
Micropterus dolomieui	A4	98.5°	9.1
	B	101.5°	9.0
	C	110.0°	9.0
	D	104.0°	7.5
	A3	99.0°	6.1
	A2	105.5°	4.8
	E	107.0°	3.1
	A1	108.5°	3.0
Micropterus salmoides	F4	94.0°	12.3
	F3	94.0°	11.0
	G	91.0°	7.2
	H	90.0°	6.4
	I	96.5°	5.7
	F2	95.5°	5.6
	F1	97.0°	5.0
	J	93.0°	4.5
	K	94.5°	3.3

correlations r_{ij} of each X_i with every other X_j. Our task is to compute the unknown coefficients $a_1, a_2, \ldots, a_n$ from some criterion implicit in the diagram.

Now imagine a unit increase in X_1, and consider the effect of this increase on Y in two different ways.

1. The expected increase in Y per unit increase in X_1 is a familiar quantity, the simple regression coefficient r_{Y1}.

2. We may partition the effect of that unit increase in X_1 as distributed over three *paths*, as follows (Figure 4.2.2):

    i) a **direct effect** a_1 upon Y;
    ii) an **indirect effect** of $r_{21}a_2$ via X_2: an increase of 1 unit in X_1 induces an expected increase of r_{21} units in X_2 which has, in turn, a direct effect a_2 upon Y for each unit increase in X_2;
    iii) an indirect effect, similarly, of $r_{31}a_3$ via X_3.

The sum of these three effects, one direct and two indirect, must result in the same expected change in Y per unit increment in X_1 that we computed in step 1. (Of course, the values predicted for Y will be different case by case.) By summing the contributions of the paths in steps (i), (ii), and (iii), we have

$$r_{Y1} = a_1 + a_2 r_{21} + a_3 r_{31} .$$

4.2 Path Analysis in Multivariate Morphometrics

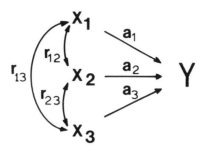

Figure 4.2.1 Multiple regression as a path analysis.

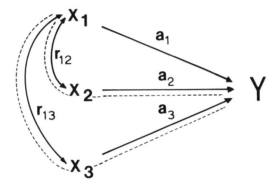

Figure 4.2.2 Direct and indirect effects.

This is one equation for our three unknowns a_1, a_2, a_3.

The partition of the effects of unit changes in X_2 and X_3 upon Y results in two similar equations

$$r_{Y2} = a_2 + a_1 r_{12} + a_3 r_{32},$$
$$r_{Y3} = a_3 + a_1 r_{13} + a_2 r_{23}.$$

These three equations are precisely the normal equations generated for minimizing $\mathrm{var}(Y - a_1 X_1 - a_2 X_2 - a_3 X_3)$ by setting to zero its derivative. In this formulation, however, we see the meaning of each equation separately as a **consistency criterion**. Each asserts the consistency between a simple regression and the net outcome of the multiple regression from the point of view of one of the X's. The solution to the whole set of equations is therefore consistent with all covariances throughout the diagram.

For an alternative exposition of this same equivalence, see Sokal and Rohlf (1981: sec. 16.3).

4.2.2 Discriminant analysis

Consider, in this spirit, the usual context of **canonical analysis** or **multiple discriminant analysis** in its simplest, two-group embodiment. We have a grouping variable G, taking on two values, and a battery of morphometric variables $X_1, X_2, \ldots, X_n$ which we wish to combine so as to separate the groups by some linear procedure. Fisher made this problem one of optimization—to find the linear combination $\Sigma d_i X_i$ for which the groups differ by the greatest multiple of their own intragroup variance. For two groups this reduces to the ordinary multiple regression of the binary variable G upon the X's. The path diagram we have just reviewed summarizes the strategy for us (Fig. 4.2.3). The **path coefficients** d_i of this diagram are the coefficients of the linear combination we seek.

But such a model is backward in this context, for it states that the canonical analysis is "predicting" a variable G that is causally *prior* to the "predictors" X_i. In fact it is G, group membership, that is causally affecting (i.e., predicting) the values of all the measures X_i, not the other way around. The proper diagram is that of Figure 4.2.4. We assume, again, that all covariances are known: both the ordinary covariances $\text{bcov}(X_i, X_j)$ in which group information is ignored (the "between" covariances), and the pooled within-group covariances $\text{wcov}(X_i, X_j)$.

How can we use the diagram to arrive at equations for the coefficients d_i? Each covariance between X_i and X_j may be partitioned into two components in this diagram: that which we observe within groups—the "within" part $\text{wcov}(X_i, X_j)$ along the two-headed arrow—and the product $d_i d_j$ contributed by the grouping, that is, by the pair of one-headed arrows in Figure 4.2.5. By analogy with the preceding case of multiple regression, we wish to partition the total covariance $\text{bcov}(X_i, X_j)$ into the separate contributions of the two paths linking these measures. Our equations become

$$\text{bcov}(X_i, X_j) = d_i d_j + \text{wcov}(X_i, X_j)$$

or

$$d_i d_j = \text{bcov}(X_i, X_j) - \text{wcov}(X_i, X_j) .$$

But from the simple algebra of cross-products we know

$$\text{bcov}(X_i, X_j) - \text{wcov}(X_i, X_j) = \Delta_i \Delta_j$$

where Δ_i is the mean difference of X_i between the groups; and thus d_i must equal Δ_i. Our discriminant function is the linear combination $\Sigma X_i \Delta_i$ in which each measure is weighted by its own mean difference between the groups.

As a practical discriminant function this expression is useless if it is mostly expressing size differences between the groups (see the discussion of the next section). (Even when we suspect the differences between the groups are mainly in size, we must check for differences of proportion anyway.) We can correct this problem immediately. A more appropriate path diagram invokes a **size-free discriminator**. The

4.2 Path Analysis in Multivariate Morphometrics

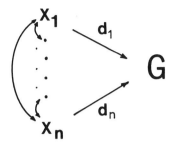

Figure 4.2.3 Path diagram for conventional discrimination.

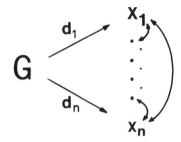

Figure 4.2.4 Corrected path analysis for discrimination.

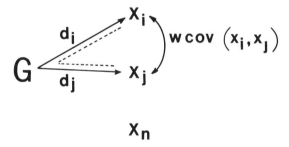

Figure 4.2.5 The two paths for a discriminant coefficient.

analysis is according to the scheme of Figure 4.2.6. Here the d_i' are the size-free coefficients we seek, S is "size" (measured as $\Sigma s_i X_i$), and ΔS is the size difference between the groups. Omission of the two-headed arrows of covariance among the X_i implies that the size factor S (computed, we hope, by the method of the next section) explains the within-group covariances among the X_i all by itself.

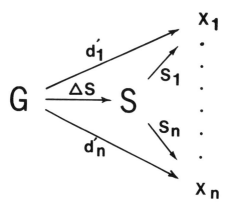

Figure 4.2.6 Path diagram for size-free discrimination.

By the principle of decomposition by paths, the coefficient we had before, Δ_i, is partitioned into a direct part d_i', to be computed, and an indirect part which is read from the diagram as $s_i \Delta S$, size difference times the effect of size on the index X_i separately. Then

$$d_i' + s_i \Delta S = \Delta_i$$

or

$$d_i' = \Delta_i - s_i \Delta S \ .$$

Each size-free discriminant coefficient is the simple predictor Δ_i corrected by subtraction of the size-based component. Our discriminant function is then

$$\Sigma d_i' X_i = \Sigma X_i \Delta_i - \Sigma s_i \Delta S X_i$$
$$= \Sigma X_i \Delta_i - S \Delta S \ .$$

In effect we have regressed S out of the size-confounded discriminator $\Sigma X_i \Delta_i$ of the preceding, unsatisfactory analysis. This regression is at the basis of the shear method for discrimination; the technique is pursued in Section 4.3. The discriminator construed by its explanatory role in this way is a much more appropriate concept for morphometric analysis than the conventional linear discriminant function which optimizes a certain variance ratio. Our purpose, after all, is not so much optimizing as explaining: what we seek in the analysis is not only description of the separation, but understanding of the roles of the variables.

4.2.3 Factor analysis

The following exposition should be considered as an exegesis of Wright (1954) or Wright (1968: v.1, chs. 13–14). It is somewhat more technical than most sections of this book, and its reading may be postponed. In this section, unlike elsewhere in this book, we follow Wright's convention, beginning the analysis with a correlation matrix rather than a covariance matrix. The factor loadings then are not allometric coefficients; rather, they are intended to express patterns of underlying genetic control. This style of factor analysis is most suited for pursuing and comparing within-group factors beyond the first. Our allometric analyses elsewhere mainly interpret small differences in coefficients upon the first and second factors only. Tukey (1954) clarifies the relation between these two purposes.

4.2.3.1 Size as an unmeasured variable.—If one measures two or more linear dimensions homologously over a period of growth of an organism, or over a population of diverse sizes, the correlation coefficients among them will almost always be very high. Each distance variable is increasing over time, and the covariance observed between pairs of such variables is due in large part to their joint dependence on growth over time. Only a small part of the coefficient is expressing functional regulation, morphology, integration, or other forms of biological constraint.

If that sort of functional explanation is one's goal the best procedure might be thought to be removing the effects of time or size from each distance variable. If the data are sampled at fixed ages, for instance, this would be equivalent to centering the observed scores at each age. Then if residual variations about the age-means are correlated one infers biological "pattern." However, organisms sampled at constant chronological age will still vary in size and, to that extent, residuals from age-corrected values will show size-confounded correlations or covariances. Standardization by *any* variable explicitly measured, whether in units of days or grams or centimeters, confounds variation of that unit with covariation among the standardized quantities.

General size explains large observed correlations in **cross-sectional data** in the same way that "general time" accounts for those in **longitudinal data**. Larger organisms should have larger distances between most landmark pairs; then in a sample having an adequate range of sizes an (unknown) part of any large correlation between linear dimensions is due to the common dependence of the measures on general size, which we almost certainly have not measured.

When we have measured more than two variables we are faced with an entire matrix of correlation coefficients r_{ij} all large and positive. It is tempting to explain the whole phenomenon at once by "joint dependence on size," but now a bit of sophistication is called for. For instance, the correlations might not be all the same; those for one variable, or one block of variables, might tend to be higher than those for others.

By invoking the algebra of factor analysis one can partial out size by constructing it from the linear measures themselves with no reference to any standard of age, length, or mass. The procedures identify size by its explanatory property: they compute an unmeasured **latent variable** in relation to which the observed correlations are *most explained.* This hypothetical variable may then be partialled out from the correlation matrix and the residual covariances examined. Scrutiny of the residuals may unearth subsets of the covariances with factors of their own. Each of these will represent a systematic pattern of morphological variation, such as growth regulation, distortion in preservation, landmark redefinition, or polymorphism, independent of the general size allometry. In the final analysis, as we will see below, a data set may manifest many factors, perhaps even many size factors, each contributing to an explanation of certain subsets of the covariances.

This will in general be our notion of **factor**: an unmeasured (latent) variable that accounts for correlations among measured variables according to a certain family of linear models, introduced below. We emphasize the distinction between factors and the **principal components** that sometimes approximate them. Principal-components analysis is not modeling but data transformation to new measures that "maximize variance" subject to certain constraints. (Rotation of components, by whatever algorithm, continues to optimize certain variances.) Principal components may be of correlation or covariance matrices. In a single population having sufficient size variation, each coefficient for the first principal component of a covariance matrix represents the **allometry** of one variable with respect to "general size" (see Sec. 2.2); the same coefficient for a component of the correlation matrix represents instead the *fraction of variance* of that variable explained by general size.

All principal components are formally the same. This is not congruent with their usual interpretation. They are extracted in order: the first explains the most variance in the data, the next the most variance among the residuals from regression-out of the first, and so on. But in fact all components simultaneously satisfy a criterion of **consistency** similar to that of Wright's first factor. Each component both predicts the indicators and is predicted by them, with the same coefficients. That is, its coefficients, term by term, are also regression coefficients of each indicator upon the score. From this characterization we can infer immediately their identity as **eigenvectors** without any need for optimization and Lagrangian multipliers. From this also follows their orthogonality, which is often inconvenient for biological interpretation (Gould et al., 1974; Atchley, 1971). Our restraint regarding principal components is not based in any distaste for optimality, but derives from the inflexibility of their definition. The biometrician is unable to adjust the second and subsequent components, even though the data may lead him to subtler models.

The method of Wright recounted here is not constrained so. Although it somewhat resembles the classic technique of oblique rotation, it differs from other algorithms currently available in that it allows the investigator interactively to vary the roles of the variables and factors in the model. Analysis in this form is not a matter of a single matrix manipulation followed by interpretation, but of cycles of interpretation, redefinition, and numerical experiment. In this sense, a factor is an explanation re-expressing the observed covariances in a particular formal framework. What makes a

4.2 Path Analysis in Multivariate Morphometrics

factor satisfactory is the adequacy of this expression for all its purposes, of which "fitting" the data is only one.

The factors we will compute come in several varieties, distinguished by their explanatory purpose: **general** and **primary** size factors, and special **secondary factors** of size, shape, or position of parts. Factors will be extracted from either covariance or correlation matrices. Although all principal components are formally the same, factors have individual identities by virtue of playing diverse roles in our models. As factors, "size" and "shape" are conceptually distinct; as components, there is guaranteed no such distinction. In multiple-group problems (Sec. 4.3) all components confound size and shape, but factor models distinguish them clearly.

4.2.3.2 How factors impute correlations.—The algebra of factor analysis can be deduced from a path diagram of the explanation it embodies. Suppose two variables X_1, X_2, each normalized (for convenience) to variance unity, manifest a correlation coefficient r which we wish to attribute to their jointly expressing a third variable Y, also of variance unity. We mean by this to assert

$$X_1 = a_1 Y + e_1, \quad X_2 = a_2 Y + e_2$$

where each of e_1 and e_2 is pure random noise uncorrelated with the other and with Y. The path diagram, Figure 4.2.7, shows one path linking the variables.

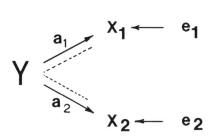

Figure 4.2.7 Path diagram for a factor on two variables.

Under these assumptions, the variance of e_i is $1 - a_i^2$ and the correlation of X_1 and X_2 is $a_1 a_2$. If their correlation is observed to be something other than this product of the separate regression coefficients on a third variable, e_1 and e_2 must be correlated. If Y is the common cause of these variables, and we wish to remove its contribution to their correlation, then the residual covariance $r_{12} - a_1 a_2$ measures the strength of functional constraint. Note that this inference proceeds from residuals of *correlations*, not variables.

If there are three variables X_1, X_2, X_3 with observed correlations r_{12}, r_{13}, r_{23}, the number of parameters to be estimated (a_1, a_2, a_3) equals the number of correlations, and the model

$$\left.\begin{array}{l}X_1 = a_1 Y + e_1 \\ X_2 = a_2 Y + e_2 \\ X_3 = a_3 Y + e_3\end{array}\right\} \quad \text{or}$$

$$Y \begin{array}{l} \overset{a_1}{\nearrow} X_1 \leftarrow e_1 \\ \overset{a_2}{\rightarrow} X_2 \leftarrow e_2 \\ \underset{a_3}{\searrow} X_3 \leftarrow e_3 \end{array}$$

has a unique solution:

$$a_1 = \sqrt{r_{12}r_{13}/r_{23}},$$

$$a_2 = \sqrt{r_{12}r_{23}/r_{13}},$$

$$a_3 = \sqrt{r_{13}r_{23}/r_{12}},$$

because, by cancellation, $a_1 a_2 = r_{12}$, $a_1 a_3 = r_{13}$, $a_2 a_3 = r_{23}$.

The model fails (i.e., the error terms e_i must be correlated) if any of the a_i are greater than one or if just one of the observed r_{ij} is negative. In the former case, some e_i would have negative variance; in the latter case, all the a_i are imaginary.

If all correlations among the X_i are equal to r, then each a_i is equal to $\sqrt{r}$. If any observed r_{ij} between two of these variables is equal to the product $r_{ik}r_{jk}$ of their separate correlations with the third variable X_k, then a_k will be 1; that is, X_k would be measuring general size perfectly. Such an assertion, which sounds odd in this terminology, means only that the diagram in Figure 4.2.8 accounts for all the observed r's, so that we need look no further for our underlying cause Y.

Each X_i is measured with noise e_i. Of the observed correlation of 1.0 between X_i and itself, only a_i^2 ($= a_i a_j$ for $i = j$) is due to the dependence of X_i on size; the rest of the 1.0 is merely the correlation of the noise in X_i with itself. In the absence of a great many more measures like X_i, we cannot actually observe this noise term separately; but we can imagine being able to measure it (by analysis of bilateral asymmetry, for instance) and so we can imagine the explained part $\hat{X}_i$ of each variable X_i. The covariance matrix of the $\hat{X}_i$ is not

$$\begin{pmatrix} 1 & r_{12} & r_{13} \\ r_{12} & 1 & r_{23} \\ r_{13} & r_{23} & 1 \end{pmatrix}$$

but

$$\begin{pmatrix} a_1^2 & r_{12} & r_{13} \\ r_{12} & a_2^2 & r_{23} \\ r_{13} & r_{23} & a_3^2 \end{pmatrix}$$

4.2 Path Analysis in Multivariate Morphometrics

—each diagonal term is diminished by the error variance of the corresponding X_i. These new diagonal terms are **communalities** or "self-correlations."

We may define Y, "general size," by formula:

$$Y = \frac{a_1 \hat{X}_1 + a_2 \hat{X}_2 + a_3 \hat{X}_3}{a_1^2 + a_2^2 + a_3^2}$$

We cannot determine its value exactly, case by case, because the error components of the X_i, which we must subtract to arrive at the $\hat{X}_i$, are unknown. But it may be algebraically verified that this unmeasureable Y has variance 1 and has covariances a_1, a_2, a_3 with $\hat{X}_1$, $\hat{X}_2$, $\hat{X}_3$. This quantity therefore estimates "general size" for three variables.

In the three-variable case the products of the a_i exactly reproduce the off-diagonal r_{ij} of the correlation matrix among the X_i. (There is no analogous set of a_i for the two-variable case—the values of the a_i are indeterminate, two parameters constrained by only a single equation.) For more than three variables we cannot guarantee in general that any set of a_i will be satisfactory. We have n a_i's with which to fit $n(n-1)/2$ r_{ij}'s, and we are therefore short $n(n-3)/2$ parameters. The question naturally arises of fitting a reasonable set of a's to the matrix. Various criteria suggested in the psychometric literature—maximum-likelihood estimation assuming a multivariate normal population, or minimizing the sum-of-squares of the residual correlations—are all more-or-less difficult to compute.

We suggest instead using Wright's criterion of **consistency**. Coefficients $a_1 \ldots a_n$ may serve as loadings for $X_1 \ldots X_n$ on "general size" if they are the covariances of the $\hat{X}_i$ with the (unmeasured) size score $a_1 \hat{X}_1 + \ldots + a_n \hat{X}_n$. In other words, for each a_i the roles of loading and path coefficient are consistent. They may not be reproducing all the correlations among the X_i, but they are at least reproducing their own. In more technical language, we require the vector $(a_1 \ldots a_n)$ to be an **eigenvector** of the correlation matrix with communalities down the diagonal:

$$\begin{pmatrix} a_1^2 & r_{12} & r_{13} & \ldots & r_{1n} \\ r_{12} & a_2^2 & r_{23} & \ldots & r_{2n} \\ . & . & . & & . \\ r_{1n} & r_{2n} & r_{3n} & \ldots & a_n^2 \end{pmatrix}.$$

4.2.3.3 Comparison with principal components analysis: Wright's Leghorn data.—This straightforward method for size-adjusting r's has been almost totally ignored in the biometric literature. Wright developed it in the 1930's as part of a general attack on problems of multiple measurement in biological explanation. (The related technique of multistage causal analysis has thrived in genetics and even sociology.) The locus classicus of his work is the analysis of six measures on a population of adult White Leghorn hens, set forth most simply in Wright (1954). Relevant tables from his article are copied here.

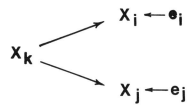

Figure 4.2.8 Path diagram for a perfect factor loading.

The data comprise six linear measurements on 276 White Leghorns: Length and Breadth of skull and lengths of Humerus, Ulna, Femur, and Tibia. The observed r_{ij} (Table 4.2.1) seem to fall neatly into two ranges, .875–.940 among the limb measures and .525–.615 between limb and head measures and between the two head measures.

An alternative analysis for data like these is principal components analysis, complete eigenanalysis of the correlation matrix with 1's in the diagonal. Table 4.2.2 presents the analysis of the data in this way. A recent biometrics text interprets each quantity of this example (Pimentel, 1979:4). We paraphrase as follows:

> The first component accounts for 76% of the (normalized) variance in these variables. As all its coefficients are large and positive, it expresses the simultaneous increase in all these variables, i.e., general growth.
>
> The second component is **bipolar**: some of its coefficients are positive and some negative, but none near zero. It is expressing an increase in skull dimensions concomitant with a decrease in all limb dimensions. This may be called skull growth at the expense of leg and bone. Whatever its nature, the differential growth it describes varies from hen to hen in the amount of 12% of the total normalized variation of dimensions.
>
> The third component, accounting for 7% of the variance, is a contrast between the dimensions of the skull alone. This contrast is uncorrelated with both the first component (expressing general growth) and the second (contrasting skull with limbs).
>
> The last three components account for a total of 5% of the original normalized variance. Component four appears to contrast wing dimensions against leg dimensions; component five mainly contrasts femur with tibia; component six (which explains the least possible variance) contrasts humerus with ulna. Each of these components is uncorrelated with all the others. The covariance matrix is thereby "explained" by one joint growth trend together with five mutually orthogonal contrasts, of varying explanatory power, each orthogonal to the general trend.

A simpler recounting of these same observed associations is possible if one allows different factors to have different roles in the model: this is Wright's method. Wright first fit a single-factor model to these data; the loadings and residuals are presented in Table 4.2.3 as a vector of six estimated a_i and a matrix of residual correlations. The vector $(a_1 \ldots a_6)$ is highly concordant with the first principal component (Table 4.2.2), but its loadings are systematically smaller (corresponding to our intention of

Table 4.2.1 Descriptive statistics for six measures of White Leghorn fowl.

Measure	Mean (mm)	S.D.	Correlations					
			L	B	H	U	F	T
Skull Length	38.8	1.26	1.	.584	.615	.601	.570	.600
Skull Breadth	29.8	3.13		1.	.576	.530	.526	.555
Humerus	74.6	2.84			1.	.940	.875	.878
Ulna	68.7	2.73				1.	.877	.886
Femur	77.3	3.20					1.	.924
Tibia	114.8	5.00						1.

Table 4.2.2 Principal components of the Leghorn correlation matrix.

	Component					
	I	II	III	IV	V	VI
Skull Length	.7426	.4536	.4922	−.0205	.0073	−.0007
Skull Breadth	.6975	.5886	−.4085	−.0008	.0021	−.0144
Humerus	.9477	−.1583	−.0259	.2182	.0472	.1623
Ulna	.9403	−.2125	.0071	.2033	−.0423	−.1654
Femur	.9287	−.2351	−.0382	−.2139	.1841	−.0323
Tibia	.9407	−.1908	−.0296	−.1950	−.1945	.0446
Variance	4.5677	.7141	.4122	.1731	.0758	.0569

Table 4.2.3 Loadings on a simple size factor, and residual correlations.

	L	B	H	U	F	T
Loadings a_i	.665	.615	.953	.942	.923	.941
			Residuals			
Skull Length	.5574	.1749	−.0188	−.0256	−.0441	−.0263
Skull Breadth		.6219	−.0098	−.0491	−.0416	−.0239
Humerus			.0924	.0428	−.0043	−.0189
Ulna				.1130	.0077	−.0007
Femur					.1480	.0550
Tibia						.1137

explaining less variance). The four limb lengths load fairly evenly and strongly on "general size"; the head measurements have smaller loadings. The entries in the residual matrix (Table 4.2.3) are computed as $r_{ij} - a_i a_j$: for instance, the .5574 at upper left is $1.0000 - .665^2$, the .1749 next to it is $.584 - .665 \times .615$. Apart from the diagonal terms, which are estimates of "noise," there are only three distinctly positive residual correlations in the matrix: those for length of skull with breadth (.1749), humerus with ulna (.0428), and femur with tibia (.0550). Note that length of humerus is the "purest" measure of size of these size variables—it has the least noise (9.2%) according to the general-size model.

To a geneticist such as Wright, the pattern of those large residual correlations is precisely in accordance with the theory of growth regulation organ by organ rather than by grand "contrasts" from end to end of the organism. He would deny an explanation of skull growth "at the expense of" leg and wing, on the grounds that there is no known genetic mechanism or developmental model for this. Instead the data are suggesting precisely the three specific genetic expressions we ought to be testing: specific factors accounting for the head measure correlation, the wing measure correlation, and the leg measure correlation. (Of course, we know nothing about these factors beyond the fact of their accounting for residual correlations.)

Wright (1954) tersely introduces an elegant variation of the method to adjust the first factor for the postulation of these secondary factors. Though he does not explain it this way, he has in fact replaced the earlier problem, of the vector $(a_1 \ldots a_6)$ as eigenvector of the matrix

$$\begin{pmatrix} a_1^2 & \ldots & r_{16} \\ & \ldots & \\ r_{16} & \ldots & a_6^2 \end{pmatrix},$$

with a new problem, that of finding a vector $(a_1 \ldots a_6)$ that is an eigenvector of

$$\begin{pmatrix} a_1^2 & a_1 a_2 & r_{13} & r_{14} & r_{15} & r_{16} \\ a_1 a_2 & a_2^2 & r_{23} & r_{24} & r_{25} & r_{26} \\ r_{13} & r_{23} & a_3^2 & a_3 a_4 & r_{35} & r_{36} \\ r_{14} & r_{24} & a_3 a_4 & a_4^2 & r_{45} & r_{46} \\ r_{15} & r_{25} & r_{35} & r_{45} & a_5^2 & a_5 a_6 \\ r_{16} & r_{26} & r_{36} & r_{46} & a_5 a_6 & a_6^2 \end{pmatrix}.$$

In this matrix, not only the diagonals but also the postulated residual factor spaces have been filled by terms exactly fitting the model, so that the actual values r_{12}, r_{34}, r_{56} in those cells are reserved for later analysis. In other words, the 2×2 blocks

$$\begin{pmatrix} a_1^2 & r_{12} \\ r_{12} & a_2^2 \end{pmatrix}, \begin{pmatrix} a_3^2 & r_{34} \\ r_{34} & a_4^2 \end{pmatrix}, \begin{pmatrix} a_5^2 & r_{56} \\ r_{56} & a_6^2 \end{pmatrix}$$

4.2 Path Analysis in Multivariate Morphometrics

have been *sequestered* from the main body of the reduced correlation matrix. We will refer to such modifications (a_i) of the general size factor as **primary size factors**, as they correspond to a slightly different model.

The residuals from this primary-factor model are presented in Table 4.2.4. The coefficients of the size factor have been somewhat altered by this respecification; the residual correlations between the head measures and the limb measures are reduced nearly to zero; and the residuals for the correlations "masked" by the modified method have mostly grown larger: .213 for .175, .072 for .055, .033 for .043. To each of these corresponds a secondary size factor representing its "explanation." (It is possible that the smallest of these residuals is now not worth its own factor, especially since the two wing measures load most strongly on general size.)

There is no way to determine path coefficients for "factors" on two variables only; they are set, by Wright's convention, at the square-root of the correlation they are to explain.

There results the explanatory scheme of Table 4.2.5, which leaves unexplained residual correlations ranging from $-.022$ to $+.017$. The largest of these is statistically significantly different from zero but, as Wright notes, not worth explaining. We have explained the observed correlations by four joint causes (Fig. 4.2.9): a primary size factor and three secondary factors.

We may compare this analysis with the conventional principal-component approach as follows. First, Wright-style analysis involves fewer parameters, 15 (Table 4.2.5) instead of 36 (Table 4.2.2). Of Wright's 15 parameters, nine are algebraically independent; of the 36 coefficients of the principal components, fifteen are algebraically independent. Likewise, the diagrammatic summary of Wright's method has only four abstract objects in it; the summary of the conventional exegesis would involve a dozen or more, at least two for each side of each "contrast." Second, the path analysis corresponds to a reasonable biometric model—joint dependence of measured variables upon hypothetical factors—whereas principal components analysis corresponds to no model at all. Third, even though the size factors found by the two techniques are nearly identical, the subsequent factors are remarkably different. In Wright's analysis, which is akin to an interactively designed oblique rotation, each factor has its own particular role in the model; whereas in principal components analysis all factors have the same formal characterization.

Statistical tests for differences between alternate factor models, or for differences of factor structure or factor loadings between groups, are the general concern of **confirmatory factor analysis**, a family of maximum-likelihood techniques for fitting the same sort of patterned factor schemes we deal with here. They require the assumptions of a multivariate normal model. Coefficients from a confirmatory factor-analysis program may be slightly different from these numbers, because they maximize a posteriori "likelihood" while we merely require consistency of the coefficients. These simpler least-squares estimators have greater robustness, while maximum-likelihood estimators are more precise when the multivariate normal model is actually true. For extended discussions of this difference see, for example, Fornell (1982) or Jöreskog and Wold (1982).

Table 4.2.4 Loadings and correlations on primary size after sequestering. Reproduced correlations are products of pairs of loadings.

	L	B	H	U	F	T
Loadings	.636	.583	.958	.947	.914	.932
		Reproduced Correlations				
Skull Length	.405	.371	.609	.603	.581	.593
Skull Breadth		.340	.559	.552	.533	.544
Humerus			.917	.907	.875	.893
Ulna				.897	.865	.883
Femur					.835	.852
Tibia						.869
		Residual Correlations				
Skull Length		.213	.006	−.002	−.011	.007
Skull Breadth			.017	−.022	−.007	.011
Humerus				.033	.000	−.015
Ulna					.012	.003
Femur						.072
Tibia						

Table 4.2.5 Path coefficients for the Leghorn model with four factors.

	Primary	Head	Wing	Leg	Error
Skull Length	.636	.461			.619
Skull Breadth	.583	.461			.669
Humerus	.958		.182		.222
Ulna	.947		.182		.265
Femur	.914			.269	.304
Tibia	.932			.269	.243

4.2.3.4 Data with a geometric structure: two examples.—We measured the elements of a truss (see Sec. 3.2): eleven distances among six midsagittal landmarks along the outline of *Cottus cognatus*, in lateral view. The sample comprised 15 specimens over a threefold size range. The numbered variables correspond to measured distances as in Figure 4.2.10.

The correlations among the logarithms of the measured distances are at the top of Table 4.2.6. The loadings on general size, just below, range from .776 to .998. Of the residuals there are large values among five of the six pairs pertaining to variables 2, 3, 7, 9. We sequestered those correlations, so that they were not estimated by $a_i a_j$, and

4.2 Path Analysis in Multivariate Morphometrics

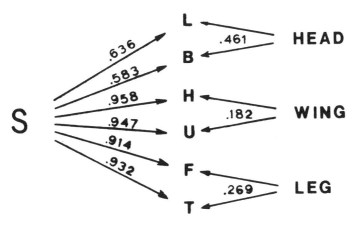

Figure 4.2.9 Path diagram for Wright's Leghorn data.

recomputed size as a primary factor rather than the general factor; the loadings on variables 2, 3, 7, 9 did not change by much. The newly unconstrained 4 × 4 residual matrix regarding variables 2, 3, 7, 9 is at the top of Table 4.2.7. A special-factor model fits these four variables quite well; its loadings are neatly patterned, positive for variables 2 and 3 and negative for variables 7 and 9. Because all these distances share a common landmark, the anterior terminus of the pelvic fin, the meaning of the special factor can be expressed kinematically: there is systematic anteroposterior variation in the position of this point, uncorrelated with the allometries of general or primary size. This may be expressing a pattern of transient hinging of the pelvic girdle (owing to feeding state, distension by eggs, or artifact of preservation) or a permanent shape polymorphism. Like size, any such factor must be explicitly controlled in the course of further taxonomic analysis. The model we arrive at requires two factors to explain the associations among variables 2, 3, 7, 9; a one-factor model leaves unacceptably large residuals for these four variables. There are also large residual correlations between variable 5 and several distances of this set—too many, in fact, suggesting no particular pattern. We were unable to model the relations of variable 5 satisfactorily by any scheme of subsequent factors.

For a second example we analyzed a collection of measures to one or another endpoint of the atherinid *Atherinella* sp.: distances to either snout or hypural plate, as shown in Figure 4.2.11. (These measures were selected from a larger data set including fin lengths, fin base lengths, and body widths and depths.) Principal component analysis of the correlations among these distances (Table 4.2.8) shows a second factor clearly contrasting the distances to the snout against the distances to the hypural plate. In a factor model this would be expressed as a secondary factor for either hypural landmark position or snout landmark position. (Measure 1, Standard Length, straddles the subsets.) The range of loadings on the second component is large; but the range of the loadings on a secondary factor is about 3:2 for the distances-to-snout, about 3:1 for the distances-to-hypural-plate. We see that the factor loadings are much

Table 4.2.6 Analysis of size factors for the 11 distances in Figure 4.2.10.

	1	2	3	4	5	6	7	8	9	10	11
						Correlations					
1	1.	.832	.799	.927	.679	.930	.698	.817	.782	.651	.899
2		1.	.976	.910	.782	.795	.664	.816	.671	.703	.817
3			1.	.873	.754	.777	.569	.778	.635	.759	.799
4				1.	.797	.933	.816	.938	.859	.782	.937
5					1.	.640	.613	.733	.660	.706	.648
6						1.	.787	.894	.913	.741	.977
7							1.	.784	.799	.591	.780
8								1.	.899	.807	.910
9									1.	.769	.905
10										1.	.774
11											1.
					Loadings, General Size Factor						
	.905	.894	.865	.998	.776	.951	.792	.947	.887	.811	.959
				Residual Correlations from General Size Factor							
1	.181	.023	.017	.024	−.023	.069	−.018	−.040	−.020	−.082	.032
2		.200	.203	.018	.088	−.056	−.044	−.031	−.123	−.022	−.040
3			.253	.010	.083	−.046	−.115	−.040	−.132	.058	−.030
4				.005	.023	−.016	.026	−.007	−.026	−.027	−.019
5					.398	−.098	−.002	−.002	−.029	.076	−.096
6						.095	.034	−.007	.069	−.030	.065
7							.373	.034	.096	−.050	.021
8								.103	.059	.039	.002
9									.213	.050	.054
10										.343	−.003
11											.081

Loadings, Primary Size Factor (Distances 2, 3, 7, 9 Sequestered)

	.902	.891	.872	.995	.774	.949	.803	.944	.915	.808	.956
Residual Correlations from Primary Size Factor											
1	.187										
2	.028	.205									
3	.013	.199	.240								
4	.030	.024	.006	.011							
5	−.019	.092	.079	.027	.401						
6	.074	−.051	−.050	−.011	−.094	.100					
7	−.026	−.052	−.131	.017	−.009	.025	.355				
8	−.035	−.026	−.045	−.001	.003	−.002	.026	.108			
9	−.043	−.145	−.163	−.051	−.049	.044	.064	.035	.162		
10	−.078	−.018	.054	−.022	.080	−.026	−.058	.044	.029	.346	
11	.037	−.035	−.034	−.014	−.092	.070	.013	.008	.030	.001	.086

Table 4.2.7 Secondary factor of four pelvic distances in Figure 4.2.10.

	2	3	7	9
	\multicolumn{4}{c}{Residual Correlations}			
2	.205	.199	−.052	−.145
3		.240	−.131	−.163
7			.355	.064
9				.162
	\multicolumn{4}{c}{Loadings, Second Factor}			
	−.370	−.552	.205	.323
	\multicolumn{4}{c}{Residual Correlations (Two Factors Removed)}			
2	.061	−.002	.026	−.019
3		.038	−.022	.012
7			.312	−.005
9				.052

more stable than the component loadings—they are not as sensitive to quirks of optimization—and therefore are more conducive to modeling in terms of variation in position of specific landmarks, allometric gradients, and the like.

In this way point-position factors may be checked at every landmark of a triangulated or truss distance network. The points having the most independent positional variation will have the strongest such factors, whether that signal be biologically meaningful variation or noise, and whether it be in one direction or two. Any factors that emerge may be dependent on size (cf. Hopkins, 1966), indicating nonlinearities of allometry along some gradient through the point. For instance, in principle there will always exist one or more factors for curvature of the allometric regression against general size. If the primary factor is restricted to the variables whose allometric coefficients tend to increase with increasing size, then a Wright-style analysis would identify a second, compensating size factor, "retardation"; if the primary factor were restricted to the variables whose allometric coefficients decrease with size, then the second factor would be "acceleration." (See Section 5.2.1.2.) There is clearly no biological difference between the resulting models.

Other kinds of systematic geometric effects may sometimes emerge as factors. In an unpublished analysis of plane forms reconstructed from parallel sections in an oblique direction, there emerged a factor contrasting all distances measured mediolaterally with all distances measured anteroposteriorly. But this was merely a *foreshortening* factor expressing noise in the anatomical orientation of the microtome.

The core of any algorithm for computing secondary factors is the replacement of general size by primary size: eigenanalysis of the correlation matrix with various diagonal blocks sequestered. A Fortran subroutine is presented in Appendix A.5.1.3.

Table 4.2.8 Analysis of size factors for distances in Figure 4.2.11. SL is Standard Length.

			Distances to Snout							Distances to Hypural Plate				
	SL	2	3	4	5	6	7	8	9	10	11	12	13	14
					Correlations									
SL	1.	.995	.994	.993	.987	.992	.991	.980	.970	.979	.990	.985	.996	.991
2		1.	.996	.991	.985	.992	.989	.980	.962	.973	.981	.980	.991	.987
3			1.	.990	.987	.992	.989	.982	.958	.972	.985	.977	.991	.987
4				1.	.987	.991	.990	.980	.965	.972	.982	.976	.994	.987
5					1.	.986	.985	.977	.958	.962	.978	.965	.986	.982
6						1.	.996	.987	.961	.967	.983	.974	.990	.987
7							1.	.985	.964	.967	.983	.973	.989	.986
8								1.	.947	.954	.972	.959	.978	.977
9									1.	.980	.970	.977	.974	.972
10										1.	.980	.988	.978	.977
11											1.	.989	.990	.988
12												1.	.984	.984
13													1.	.992
14														1.
					Coefficients of First Two Principal Components									
I	.269	.268	.268	.268	.267	.268	.268	.266	.264	.265	.268	.266	.269	.268
II	.046	.143	.186	.135	.229	.227	.197	.356	−.525	−.471	−.133	−.373	.000	−.031
				Loadings for Model with Primary Size Factor and Snout Position Factor										
Size	1.000	.990	.989	.990	.983	.988	.988	.975	.976	.984	.994	.989	.999	.996
Snout		.105	.114	.097	.118	.138	.127	.157	.182	.196	.101	.176	.056	.069
				Loadings for Model with Primary Size Factor and Hypural Position Factor										
Size	1.000	.996	.996	.996	.990	.996	.995	.984	.967	.974	.988	.979	.995	.991
Hyp.		.105	.114	.097	.118	.138	.127	.157	.182	.196	.101	.176	.056	.069

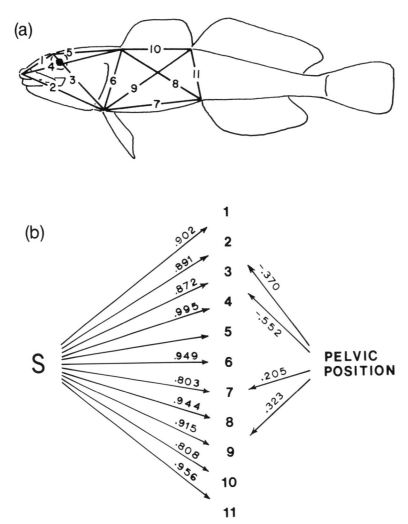

Figure 4.2.10 Eleven distance measures for *Cottus*. (a) The distances express the form of two truss cells in the anterior of the fish. (b) The findings of the path analysis: a primary size factor and a secondary factor for position of the anterior terminus of the pelvic fin.

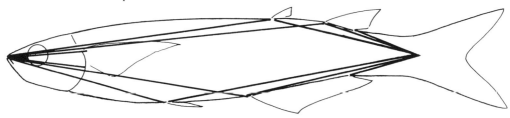

Figure 4.2.11 Distances to endpoints of *Atherinella*.

4.2.4 Concluding remarks

In our view, versions of Sewall Wright's general technique of path analysis are the only meaningful *models* available for the analysis of multivariate morphometric data.† Our basic explanatory variables—"size" and "shape," mainly—are, after all, unmeasured; they need to be formally defined and computed in terms of their properties in an appropriate model from this class. They are to be defined, in other words, by the manner in which they explain observed covariances. Before proceeding with any systematic form-comparisons between groups, it is essential to explore the within-group factor structure, so as to identify factors of shape variation, or accidents of measurement pattern, which would bias group comparison to an unknown extent. Whether our intention is to rationalize them or merely to constrain them, size and shape are explanations well before they are variables.

In application to multiple regression, path analysis yields a quite different interpretation of the normal equations for the usual coefficients. In application to discriminant analysis, path analysis indicates the incongruity of the usual canonical procedure and leads us to an algorithm for loadings that should not much resemble the conventional Fisherian coefficients. In the application to factor analysis, the characterization of principal components as combinations explaining the most variance is replaced by a criterion of explaining *co*variance instead. The resulting class of causal models, which is hypothesis-driven and interactive, is far more subtle and flexible than ordinary eigenanalysis for components after the first.

4.3 The Shear: Size-Free Discrimination

We need procedures for discriminating among groups of organisms that vary in size. The groups included in a study can be chosen *a priori* (e.g., several species or geographic populations within a species) or *a posteriori* (as a conclusion resulting from some method of analysis). However the groups are chosen, it has long been considered desirable to discriminate among them on the basis of size-free shape derived from distance measures. We construe size and shape not as measured variables, but as general factors, linear combinations most parsimoniously accounting for the associations among the distance measures (Sec. 4.2). Size, in particular, is not a single variable such as biomass or a standard length, but a factor which, when called upon to predict all the distance measures within a population, leaves the smallest mean-squared error. The quantification of size as a general factor depends on the group under study; a measure of shape depends on the differences among the groups. In the

---

† There is a large literature of other extensions of path analysis, usually under the rubric of "causal modeling of unmeasured variables" in the social sciences. Good sources for these materials include the edited proceedings of Jöreskog and Wold (1982) and the selections edited by Fornell (1982).

preceding section we referred to secondary size factors as aspects of shape. In this section our shape descriptors will always be shape discriminators defined in the context of multiple groups.

Three classes of methods have commonly been used to compare shape among groups while removing size: ratios, regressions, and factor or component analyses. The disadvantages of ratios are discussed in Section 2.2 and a reappraisal of regression and discriminant functions is presented in Section 4.2. A third class of multivariate methods, factor or component analyses, avoids such conceptual or statistical difficulty. Their coefficients, being ordinary regression coefficients of the indicators on the factors, are more easily interpretable (Stroud, 1953). In morphometric studies of groups varying in size, the first factor or component has often been interpreted as "size," and subsequent factors designated as "shape." However, Mosimann (1970), Sprent (1972), and Mosimann and James (1979) have pointed out that this labeling of component I as size and subsequent components as shape is rather arbitrary (see Sec. 2.2). They also state that the conceptual independence of these shape and size components is defined only by the orthogonality of the components, and that components subsequent to the first contain both shape and size. For the purposes of discrimination, it is size change and shape change which should be conceptually orthogonal. To this end we need to modify the components by a method we call the **shear**. Like principal component II, any discrimination-bearing component confounded with size may be sheared. For example, a meristic component heavily influenced by number of gill rakers (frequently a size-related variable) would be a likely candidate for shearing.

4.3.1 Discrimination as a component analysis

When multivariate populations are fairly cleanly separable they can be differentiated by a variety of linear discriminant functions. Although the traditional linear discriminator of Fisher is optimum for a certain variance-ratio criterion, an infinity of other linear functions may classify specimens quite as well. Of these equivalently accurate formulations we wish to approach a linear function the coefficients of which are, in some sense, the most meaningful. A coefficient has meaning, in this context, to the extent that it may serve as a **loading**, a part-whole covariance—the covariance of the variable with the linear function to which it contributes (Sec. 4.2). We have previously characterized both component and factor analyses in terms of this **consistency criterion**. In factor analysis, the covariance matrix is formally partitioned into one portion representing dependence of the variables on joint antecedent factors and another portion representing pure regression error. In this case the factor score cannot be computed, only estimated, from its observed regressands; from this dilemma follow many of the complexities and obscurities of the factor-analytic literature. For our purpose here, which is discrimination, the fine points of factor definition may be ignored. There remains the idea of the loading, the part-whole covariance expressing

4.3 The Shear: Size-Free Discrimination

the way in which a component explains variables at the same time that it is explained by (composed of) them.

The principal components for our two-group system are not formally interchangeable. One component, size, grows within individuals; the other, shape, is a record of intergroup differences. The components we seek have different names, and different hypotheses account for their patterns of loadings (Fig. 4.2.6). We may augment the application of component models in morphometrics, then, by delineating distinct criteria for these distinct components. In particular, we may specify that for each component the explanation of covariance in the context of the consistency criterion be controlled for another variable—that the loadings be partial, not total, covariances.

Consider Figure 4.3.7b, a scatter of the first two ordinary principal components for a pooled sample of two clearly distinguishable groups. Within and between groups, each of components I and II overlaps notions of both size and shape. We wish to separate these cleanly in the context of discrimination, and in particular ensure that our intergroup discriminator is not confounded by growth within groups. The "optimal" linear discriminator is, of course, confounded with size differences, while the simple partial discriminator of Burnaby (1966) has coefficients that are not loadings and that can therefore not be compared among themselves (Humphries et al., 1981).

The characterizations which suit our analysis in the spirit of Section 2.2 are as follows. In the presence of groups we define **within-group size** as a component whose loadings are covariances with group held constant: that is, a linear combination whose coefficients are its own pooled covariances within group. The loadings on this component, then, are computed from the group-free (centered) variables. This new size factor, S, which is correlated with group, corresponds to the horizontal axis of Figure 4.3.7c. If we now partial out size within groups, but leave the group means in place, we arrive at the variables with respect to which shape difference is to be a component. As a component, it will be much more stable under substitution of variables than the multiple regression coefficients with group as a dependent variable—the usual two-group discriminant function (Sec. 4.2.2). Formally, our shape discriminator, H, is a linear combination whose coefficients are equal to its partial covariances with the log-distance measures controlled for within-group size. Because the groups may differ in size or shape, our size factor is correlated with group, while shape may be correlated with size across group.

In ordinary principal component analysis (in which loadings equal covariances), to get component II one usually partials component I out of all variables and recomputes the "first" component. In the modified procedure (loadings equal to covariances controlling for group), we partial group out of all variables and again compute the usual first principal component. When computing component scores we must use the original log-transformed variables. (This distinction does not matter in ordinary principal component analysis because successive components are orthogonal.)

A simplification is possible when intragroup size is nearly within the plane of the first two ordinary principal components of the pooled covariance matrix **Q**. In this case size and shape discrimination together are nearly equivalent to the unmodified first two components together, so that the factor for shape, H, is approximated by the residual of component II after regressing out intragroup size. Geometrically, the effect is of **shearing** so that intragroup size, S, is transformed to the horizontal, while the orientation of the vertical is left unchanged at constant size (Fig. 4.3.1). Should that appear to be all the discrimination in the data, that is, should the third and subsequent components of **Q** show negligible group differences, then this residual is very nearly the H of the factor characterization above, the first principal component of the group-adjusted and size-adjusted distances.

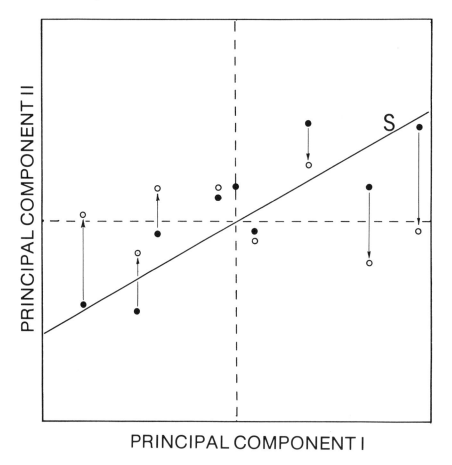

Figure 4.3.1 The geometric effect of shearing a scatter. Intragroup size is made horizontal; vertical distances are unchanged at constant size.

4.3 The Shear: Size-Free Discrimination

Had H not been constrained to be uncorrelated with S within group, then it and S would be the ordinary first two principal components. But by this constraint, H will bear all of the size-free discriminatory power among the groups. Because H is uncorrelated with size within group, we call it a shape discriminator. Unlike the explicit shape vectors of Mosimann (1970) and others, our shape discriminator H has coefficients that do not necessarily sum to zero. That is, two organisms identical except for change of physical scale will have different scores on H if the mean shapes at constant size are different for their two taxa.

The analysis we propose ignores any divergence among the principal axes of the population groups taken separately. We have discussed this difficulty in Humphries et al. (1981). In most examples, the angles between within-group size axes are small. Because there is no difficulty in interpreting the sheared components that result, we ignore these slight divergences.

4.3.2 Calculation of a sheared component

To this point we have discussed the rationale for calculation of separate size and shape factors S and H, but have not indicated how to obtain them. Computing the sheared principal component involves several steps: transformations of measured variables, eigenanalyses of unstandardized matrices, multiple regressions, and linear combinations of factor scores. For exploratory work we suggest an interactive analysis package such as *MIDAS* (see Appendix A.5), but a batch-oriented system capable of retaining intermediate matrices will perhaps suffice. An implementation of a typical shear in *SAS*, for instance, is presented in Appendix A.5.1.2.

To demonstrate the algebra of the method, we will work an example in full detail and with excessive precision.

1. The data for this example are length L and width W measurements, partly contrived, from the Weberian apparatus of three specimens each of two species of catfish from the genus *Ictalurus*.

Species	Specimen	Length(mm)	Width(mm)
1	1	15.0	16.4
1	2	16.6	19.2
1	3	18.0	22.1
2	4	18.6	15.3
2	5	22.0	22.0
2	6	24.0	27.3

2. The data are transformed to logarithms to any convenient base, in this example, base 10.

3. The total (pooled among-group) covariance matrix, **Q**, is calculated for the logs of the six bivariate observations.

$$\mathbf{Q} = \begin{pmatrix} \text{var}(\log L) & \text{cov}(\log L, \log W) \\ \text{cov}(\log L, \log W) & \text{var}(\log W) \end{pmatrix} = \begin{pmatrix} .00517 & .00526 \\ .00526 & .00863 \end{pmatrix}$$

4. The first two principal components (axes, eigenvectors) of this matrix are computed in the usual fashion, producing a matrix of column eigenvectors, **E**. Note that these must be components of the covariance matrix, not the correlation matrix.

$$\mathbf{E} = \begin{pmatrix} e_{11} & e_{12} \\ e_{21} & e_{22} \end{pmatrix} = \begin{pmatrix} .607 & -.794 \\ .794 & .607 \end{pmatrix}$$

Here e_{11} refers to the loading for the first variable on the first eigenvector, e_{12} the first variable on the second eigenvector, and so on. We have chosen to scale the elements of the matrix so that it is orthogonal; in particular, the sum of squares of each column equals 1.0.

5. Projected scores for the kth specimen on the jth component are computed from the formula

$$y_{jk} = e_{1j}d_{1k} + e_{2j}d_{2k}$$

where the e's are as above and d_{ik} is the ith log-distance variable for the kth individual. This produces a score for each individual on each component of the new coordinate system. For example, the score for the second individual on the first component is $y_{12} = e_{11}d_{12} + e_{21}d_{22}$.

Specimen k	y_{1k} (PCI)	y_{2k} (PCII)
1	1.679	0.197
2	1.760	0.190
3	1.830	0.181
4	1.712	0.289
5	1.882	0.251
6	1.979	0.225

4.3 The Shear: Size-Free Discrimination

A more convenient formula for the calculation of scores on H is:

$$H = \text{PCI}(-\alpha\beta_1) + \text{PCII}(1 - \alpha\beta_2)$$

$$= \text{PCI}(.204 \times .979) + \text{PCII}(1 - .204 \times .204)$$

Scores for each of the six specimens are calculated according to this formula.

Specimen	Sheared PCII (H)
1	.5241
2	.5340
3	.5392
4	.6194
5	.6170
6	.6107

11. The loadings for each of the variables on this new factor are computed by the same formula as was used to compute factor scores.

$$h_L = e_{11}(-\alpha\beta_1) + e_{12}(1 - \alpha\beta_2)$$

$$= .607(.204 \times .979) - .794(1 - .204 \times .204) = -.640$$

$$h_W = .794(.204 \times .979) + .607(1 - .204 \times .204) = .740$$

where the e's are as in step 4 and the h's refer to the loadings on H.

Scores on this new factor can be plotted against scores on S or PCI, or some other variable of interest. A plot of H against PCI verifies that indeed the discrimination is size-free.

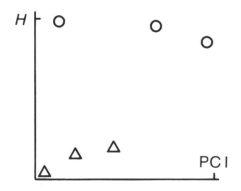

4.3.3 The role of secondary factors in discrimination

To illustrate the robustness of the shear method we have executed a set of simulations gradually increasing the strength of within-group factors in an example of two-group discrimination. We will show, by way of contrast, that even moderate complexity in a data set makes impossible the interpretation of coefficients based on residuals from any measured variable.

Figure 4.3.2 is a path diagram for a hypothetical set of nine morphometric variables. We used this as a model to generate five simulations, for two populations of 100 individuals each. The covariances among the nine variables are accounted for by primary size S, group G and various intragroup shape factors $F_1 \ldots F_4$. As simulated, the metric variables load equally strongly on S; there are no size allometries here. (A within-group principal component analysis of their covariances would result in equal loadings for all variables.) One variable, the first, is assigned a loading of 0.0 on group—no mean difference between the groups—so that we may use it as a proxy for primary size in the conventional analyses. The loadings on G for the other eight variables were assigned one of two distinct values. Because primary size and group are latent variables, they have no natural scale. We assigned G the values 0 and 1 only; S is distributed uniformly over the range 0–2, 0–3, or 2–4, depending on the group and the simulation. The loadings on S were uniformly 50.0 and those on G 0.0, +8.0 or −8.0. Group differences on individual variables are therefore about eight units in a simulation of size range 100 units within group; and differences between variables having opposite loadings on G will be about 16 units at constant primary size.

Primary size, then, accounts for about 2500 units of covariance between any pair of variables, and group accounts for either +64 or −64 units of covariance.

Additional covariance between variables was generated for the simulations by various shape factors uncorrelated with either size or group. For each variable, unique variance (as error) was added as well. The intragroup factors account for covariance in amounts ranging from 1 to ±20 units. In simulation II, for instance, a within-group factor was generated by adding a normal deviate of variance 4.0 to each of three "observed" variables. In simulations III–V, the within-group factor structure is added as diagrammed in Figure 4.3.2. These covariances violate all the models by which shape discriminators are generally computed, therefore making the simulations more realistic.

The five simulations progress from an unreasonably simple intragroup factor structure—unique variance only—to a typically complicated one. We effect this by sequentially adding within-group factors to the postulated covariances and by adjusting the size ranges. For each example we present four types of analyses: ordinary principal components analysis, PCA of size-adjusted residuals, shear analysis, and linear discriminant analysis (also of size-adjusted residuals). We then examine the effect of intragroup factors upon the resulting ordinations and upon the computed coefficients. Our primary concern is with the coefficients, because it is discrimination by shape factors, not mere discrimination of groups, that is our primary goal.

4.3 The Shear: Size-Free Discrimination 111

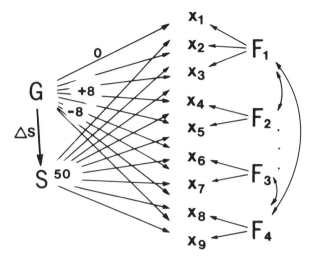

Figure 4.3.2 Path diagram for a set of five simulations. $X_1 \ldots X_9$, simulated morphometric variables. G, group shape-difference factor; S, primary size factor; $F_1 \ldots F_4$, up to four intragroup shape factors differing from simulation to simulation. Loadings: ΔS, mean size difference between groups, differing from simulation to simulation; $0, +8, -8$, loadings on shape-difference as written on arrows out of G; all X_i load 50.0 on S. Each X_i has an additional unique variance of 1.0, not shown.

Simulation I (Table 4.3.1, Fig. 4.3.3a)

In this example there are no intragroup factors, only size isometry, group difference, and unique variance of 1 unit for each variable. The distribution of S for both groups in this and the next two simulations is uniform over the range 0–2. Because the groups have the same distribution on primary size, the second principal component does not need to be sheared. All four methods produce satisfactory coefficients: two blocks of four variables with opposite signs and approximately equal magnitudes. The ordinations show the equivalent power of discrimination for each method (Fig. 4.3.3a). The mild correlation with size within group is due to random variation in the size distribution.

Simulation II (Table 4.3.1)

The addition of a single, small secondary factor to the data has only a minor effect on the two principal component methods. The coefficients of the discriminant analysis still show the expected pattern of loadings, but the individual values show considerable variation. The ordination for each method is unchanged from the previous example, but with increased within-group scatter.

Simulation III (Table 4.3.1, Fig. 4.3.3b)

The addition of four secondary factors to the original data structure has little effect on the coefficients of the two component methods. However, the coefficients of

Table 4.3.1 Simulations of group discrimination with identical size distributions.

	Simulation I No within-group factors			Simulation II One within-group factor				Simulation III Four within-group factors			
X_i	Ordinary PC II	Residuals PC I	Residuals Disc. Fn.	Ordinary PC II	Residuals PC I	Residuals Disc. Fn.		Ordinary PC II	Residuals PC I	Residuals Disc. Fn.	
1	.00	—	—	.02	—	—		.02	—	—	
2	−.35	.35	1.3	.36	.34	−.31		.36	.32	2.2	
3	−.35	.35	1.3	.37	.35	−.16		.36	.32	3.0	
4	−.35	.36	0.8	.34	.32	−.71		.34	.27	−0.2	
5	−.35	.35	1.2	.34	.32	−.73		.34	.28	0.3	
6	.35	−.35	−1.0	−.35	−.37	.42		−.38	−.44	0.1	
7	.35	−.35	−1.2	−.35	−.37	.36		−.38	−.44	−0.1	
8	.35	−.35	−1.4	−.35	−.37	.34		−.33	−.36	0.4	
9	.36	−.35	−1.0	−.35	−.37	.31		−.33	−.36	−0.4	

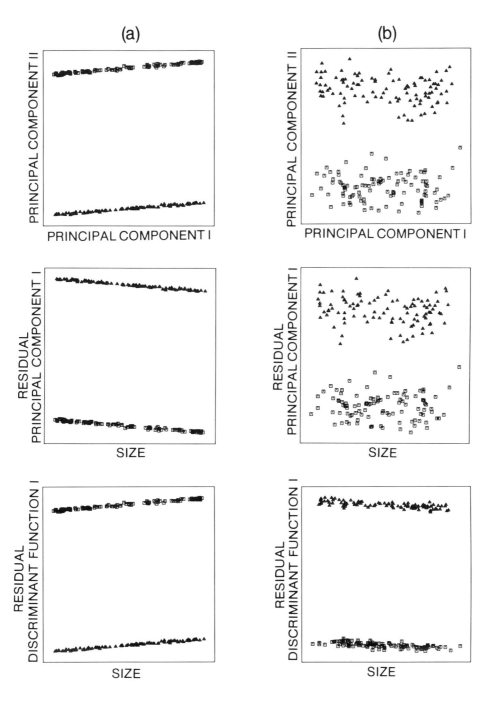

Figure 4.3.3 Scatters from simulations of group discrimination with identical size distributions. (a) Simulation I (Simulation II results are similar); (b) Simulation III. Upper row, ordinary PC II against ordinary PC I; middle row, PC I of residuals upon X_1 against X_1; bottom row, residual-based discriminant function against X_1.

the discriminant function no longer resemble the expected loadings—they no longer are interpretable. The scatter plot of components I and II from the ordinary analysis, as well as the scatter of the discriminant function against size, are as in the previous simulation (but again with increased within-group scatter).

Simulation IV (Table 4.3.2, Fig. 4.3.4)

In the fourth simulation we have superimposed a group difference in size distribution on the previous data structure. One group has size range 0–100 units, the other 0–150 units; distributions are uniform within the range of each group. Because of this it is necessary to shear the ordinary second component to estimate a size-free factor. The coefficients of the sheared component as well as the first component of the residuals still show an interpretable pattern of loadings. The sheared component results in a satisfactory discriminator, but the two residual-based analyses produce discriminators that are confounded with size (Fig. 4.3.4). It is ironic that the factor that we intended to regress out is now confounded with the first component of its own residuals.†

Simulation V (Table 4.3.2, Fig. 4.3.5)

In this simulation we reuse the data structure having four secondary factors for groups that now show little overlap in size distribution: 100–200 units for one group, 0–150 for the other. This combination adversely affects the interpretability of the coefficients in all four methods. The shear analysis produces the pattern closest to the expected result; however, coefficients that should be equal vary in a ratio of 2:1. The first component of the residuals no longer retains the basic pattern of four positive and four negative loadings.

Only the shear method produces a correct ordination—the discriminators from the other two methods are seriously correlated with size. Examples such as this last simulation illustrate the necessity of examining within-group factor structure in problems of discrimination.

---

† The residuals from primary size (Sec. 4.2.3) are, on the average, uncorrelated. But we are regressing out not primary size, but its observed proxy, variable one—primary size plus some unique variance. Thus, the residuals of the other variables on this particular size measure each share some multiple of that unique variance. This produces another factor in the data structure, one correlated with size. In the first three simulations this factor is sufficiently small that it does not influence discrimination. But in the last two simulations, where discrimination is confounded by both size and within-group factor structures, this factor has a noticeable effect on the two residual-based discriminators. It should be noted that this is not a result of our method of producing the variables. Whichever variable in a real data set is used for size will have a comparable variance term, leading to a comparable covariance.

4.3 The Shear: Size-Free Discrimination

Table 4.3.2 Simulations of group discrimination with four intragroup size factors and differing size distributions.

X_i	Simulation IV Little overlap				Simulation V Much overlap			
	Ordinary PC II	Sheared PC II	Residuals PC I	Residuals Disc. Fn.	Ordinary PC II	Sheared PC II	Residuals PC I	Residuals Disc. Fn.
1	.02	.03	—	—	.10	.12	—	—
2	.35	.36	.33	.43	.38	.40	−.16	.23
3	.35	.36	.33	.29	.38	.40	−.16	.25
4	.32	.33	.28	−.01	.20	.22	.11	−.01
5	.32	.33	.28	.03	.20	.22	.11	.02
6	−.37	−.36	−.41	−.05	−.50	−.48	.62	.03
7	−.37	−.36	−.41	.04	−.50	−.48	.62	−.04
8	−.37	−.36	−.38	−.01	−.24	−.22	.27	.01
9	−.38	−.36	−.39	.02	−.24	−.22	.27	−.02

4.3.4 Use of the shear method to identify hybrids and unknowns

Three examples of the shear analysis in shape discrimination were published by Humphries et al. (1981). In this book we have selected two others to demonstrate (1) separation of size and shape in the presence of strong within-group correlation of components I and II, and in the study of possible hybrids, and (2) identification of a cluster of small specimens of questionable genetic provenance. The examples are drawn from studies of three species in the *Gila robusta* complex from the Colorado River in the western United States. Figure 4.3.6 shows shapes of typical small and large individuals of the three species. At a length of about 200 mm *Gila cypha* develops an abrupt elevation in the profile where large muscles and tendons attach to the skull behind the dorsal surface of the head. *Gila elegans* develops similar profile changes, which are not as extreme or as abrupt as in *cypha*. *Gila robusta* retains a more typical fish shape until, at large sizes, it develops a profile that may be confused with that of *elegans*. The discrimination problem is further complicated by hybridization which produces intermediate specimens.

4.3.4.1 Intermediacy due to hybridization.—The shear method helps distinguish between intermediacy due to growth stage and intermediacy due to hybridization. Figures 4.3.7a–d show the sequential separation of size and meristic effects in the calculation of a shape discriminator among these three species. Figure 4.3.7a illustrates the scatter of scores on components calculated in the usual fashion from the correlation matrix of 33 morphometric and meristic characters. In using the mixed data set in this way, size, meristic variation, and shape variation are confounded in both principal components. The first step toward clarification is separation of the

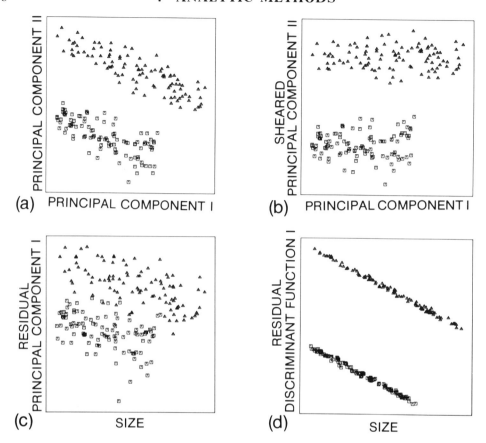

Figure 4.3.4 Scatters from Simulation IV. (a) Ordinary PC II against ordinary PC I; (b) Sheared PC II against ordinary PC I; (c) PC I of residuals from X_1 against X_1; (d) Residual-based discriminant functions against X_1.

seven meristic variables for later analysis. The 26 morphometric variables are then factored using their covariance matrix. The orientation of the clusters in Figure 4.3.7b shows that components I and II have nearly equal fractions of variance explained by size within group. Large specimens are at the upper right of each growth oval; suspected hybrids lie between the clusters. Our next step is to calculate a new component II that is uncorrelated with size—the sheared component II. The scatter of sheared component II by among-group size (Fig. 4.3.7c) does a reasonable job of discriminating the three taxa and also clarifies the position of possible hybrids with respect to the species clusters. We restore the meristic information by way of its first principal component, scattered against sheared morphometric component II (Fig. 4.3.7d). The alignment of the clusters indicates similar patterns of discrimination by the morphometric and meristic variables and calls our attention to additional possible hybrids between the clusters. Ordination by two data sets enhances discrimination of the taxa and helps define the nature of the intermediacy.

4.3 The Shear: Size-Free Discrimination

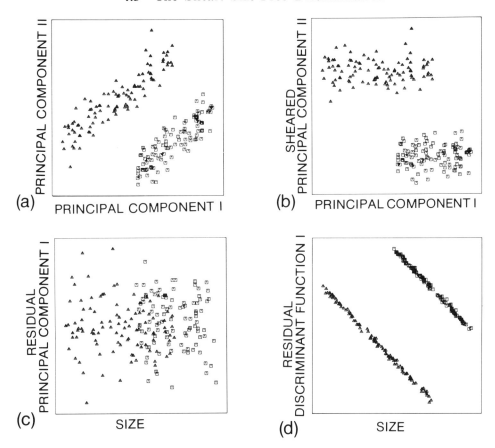

Figure 4.3.5 Scatters from Simulation V. Frames are as in Figure 4.3.4.

All of the components referred to in the example are shown in Table 4.3.3. As we have seen, in a typical principal component analysis, size and shape are confounded in the first two components. The degree to which this confounding occurs is generally due more to the extent to which size differences are reflected on principal component II than shape differences on principal component I, because specimens of different sizes will usually differ in shape anyway. In this example, the phenomenon is more pronounced than usual, in that covariation with size is nearly evenly distributed between the first two components. Because of this, the shear produces a much larger effect on both the scatter plot and the coefficients (Table 4.3.3, Figs. 4.3.7a,b). For example, the rank of the loading for caudal peduncle depth increases from seventh to third, more accurately reflecting its role as a powerful discriminator among these species. Comparing Figures 4.3.7c,d, we can see that the *elegans* cluster has changed its shape. With size removed from principal component II, the vertical dispersion in Figures 4.3.7c,d is due entirely to variation in shape.

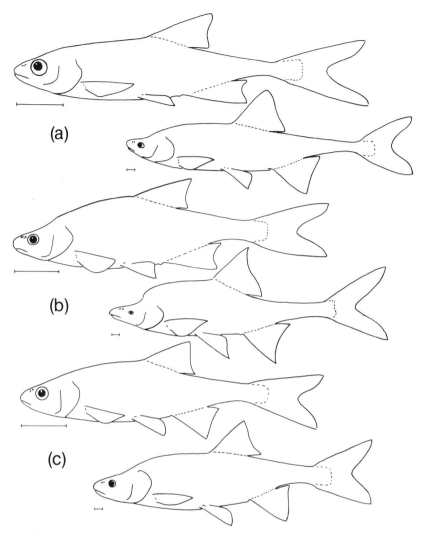

Figure 4.3.6 Shapes of young and adult members of three species of *Gila*. (a) *G. elegans*, (b) *G. cypha*, (c) *G. robusta*. Scale bar = 1 cm.

4.3.4.2 A model for hybrid recognition.—The classification of unusual specimens such as suspected hybrids is an example of the problem of possible membership in an unknown group. Rao (1973:577–79) describes the analysis of an individual's membership in one of two known groups or in a third, unknown, group. In this context, hybrids are individuals the parents of which belonged to two different groups, known or unknown. Because combination of different genomes may yield high variance, the population of possible hybrids between two groups is a third, highly variable group, between and overlapping the parental groups. The problem is compounded when unknown parental groups might be involved.

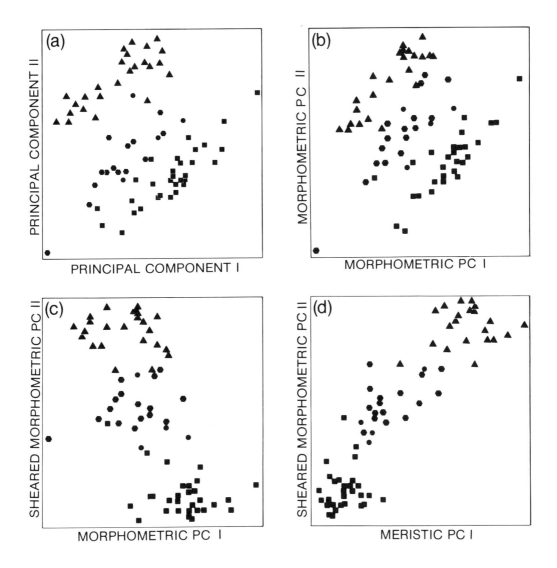

Figure 4.3.7 Principal component analyses of three species of *Gila* and their hybrids. *G. robusta* (squares); *cypha* (hexagons); *elegans* (triangles); hybrids (circles). (a) Scatter of components I and II of mixed morphometric and meristic data. (b) Diagonal clusters for morphometric principal components I and II illustrate their nearly equivalent loadings on size. (c) After shearing, the morphometric components show separate effects of size and shape on the clusters. (d) Scatter of sheared component II against meristic component I. Discrimination is enhanced. (Data supplied by Robert R. Miller.)

Table 4.3.3 Coefficients for analyses of species of *Gila*.

Variable	Three Species and Hybrids								Two Species and Unknowns		
	Pooled Variables		Morphometrics Only			Meristics Only		Morphometrics		
					Sheared					Sheared
	I	II	I	II	II	I	II	I	II	II
Standard length	.15	.30	.10	.27	.18			.22	−.19	−.20
Head length	.25	.02	.21	.06	−.04			.20	−.02	−.03
Eye diameter	.19	.05	.15	.09	.01			.11	.24	.24
Snout length	.22	−.07	.27	−.06	−.15			.21	−.12	−.13
Preanal length	.19	.13	.15	.15	.06			.22	−.06	−.07
Head depth	.22	−.15	.29	−.17	−.25			.18	.30	.29
Occiput depth	.22	.04	.18	.06	−.02			.19	.11	.10
Interorbital w.	.20	.13	.13	.12	.05			.21	−.13	−.14
Snout-occiput l.	.24	−.04	.25	.00	−.10			.20	.20	.20
Dorsal base l.	.01	.35	−.01	.40	.33			.22	−.35	−.36
Anal base length	−.03	.35	−.05	.49	.42			.22	−.46	−.47
Predorsal length	.22	.13	.16	−.14	−.18			.21	−.08	−.09
Pectoral length	.10	.19	.07	.16	.10			.20	−.27	−.28
Pelvic length	.03	.16	.01	.14	.11			.20	−.22	−.23
Upper jaw length	.22	−.09	.32	−.09	−.20			.20	.12	.11
Mouth width	.18	−.01	.19	.01	−.07			.21	.03	.03
D to P2 origin	.10	.16	.07	.15	.09			.22	−.13	−.14
Caudal ped. depth	.20	−.17	.33	−.25	−.33			.19	.16	.15
L. pharyngeal l.	.25	−.01	.22	.03	−.07			.18	.19	.18
R. pharyngeal l.	.07	.26	.06	.34	.25			.18	.14	.13
L. pharyngeal w.	.12	.24	.10	.31	.21			.18	.12	.11
R. pharyngeal w.	.07	.26	.06	.34	.25			.18	.14	.13
L. post. phar. l.	.21	.15	.14	.16	.07			.18	.11	.10
R. post. phar. l.	.21	.16	.14	.17	.08			.18	.12	.11
L. ant. phar. l.	.23	−.10	.32	−.10	−.21			.19	.24	.24
R. ant. phar. l.	.23	−.10	.29	−.09	−.20			.20	.15	.14
Dorsal rays	−.11	.25				.49	−.00			
Anal rays	−.16	.24				.47	.31			
Pelvic rays	.01	.06				.02	.25			
Pectoral rays	−.07	−.01				−.01	.73			
Lateral line sc.	−.00	−.04				.14	−.53			
Gill rakers	−.12	−.27				.51	−.15			
Vertebrae	−.13	.26				.51	.01			

4.3 The Shear: Size-Free Discrimination 121

Consider a putative hybrid and two possible parental groups. The classification problem can be stated as tests of two hypotheses: (1) that the individual belongs to one of the known groups; (2) that the individual is of mixed parentage involving two different known groups. Because one cannot attach prior probabilities to membership in unknown groups, the third possibility—parentage unknown—cannot be tested.

We suggest a heuristic model in which we calculate an ordination with variance partitioned into a within-group size factor and a group shape-difference factor (each of which contributes to the morphometric variates), and also calculate a meristic group difference factor (which contributes to the meristic variates). The shape-difference factor is uncorrelated with the size factor within groups. When the morphometric and meristic factors separately but similarly define group differences, scatter of sheared morphometric component II against a principal meristic component produces a diagonal scatter with clusters at the poles. The greater variance of hybrids is presumed to be orthogonal to this diagonal in all directions within the original high-dimensional measurement space; in this projection onto a single plane, the additional variance will be attenuated nearly to zero. In effect, we have modeled hybridization as producing additional unique variance (Sec. 4.2.3) without altering the covariance structure explained by ordination. A group of "hybrids" affecting the covariance structure, such as a hybrid population that breeds true, is not distinguishable from an arbitrary third group.

The clusters mark outer limits of a hypervolume whose length spans the two groups and whose width is not less than the variation of the groups around the line connecting their centroids. Hybrids between the known groups often fall between the clusters, but may fall in the parental clusters or even outside the hypervolume. Quantitative empirical work is needed to define better the implications of the model. Natural fish hybrids are abundant and interesting expressions of novel genomic combinations.

In this example (see Fig. 4.3.7), several hybrids identified prior to the analysis are shown by circles. The scatters in Figs. 4.3.7c,d indicate that the prior identifications were partly incorrect. If the circles do indeed represent hybrids, there are about twice as many hybrids not so identified in the sample. In particular, scatter of the specimens identified as *cypha* shows large variance not attributable to size, and probably includes numerous hybrids with the other two species.

4.3.4.3 Identification of unknowns.—A different sample of *Gila*, taken several hundred miles downstream, leads to a different analysis, demonstrating identification of a cluster of small individuals of questionable provenance. Several clusters are apparent in the scatter of principal components I and II (from the covariance matrix of the 26 log-transformed distance measures). The investigators had reason to believe that the small specimens plotted as round symbols are *Gila robusta*, whose large individuals are represented in the scatter by square symbols, rather than *Gila cypha*, the hexagonal symbols (Fig. 4.3.8). To test this, we computed a pooled within-cluster size factor from the two clusters of large specimens, then sheared so that size became

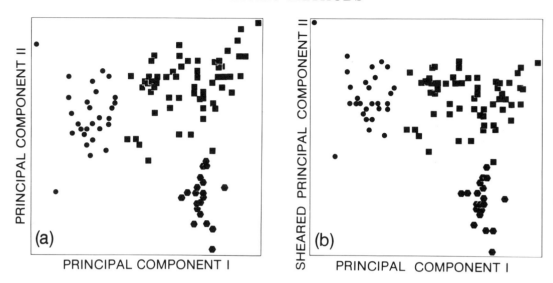

Figure 4.3.8 Identification of unknowns. (a) The two clusters represented by circles and squares may be small and large members of *Gila robusta*. (b) After shearing, the scatter supports the hypothesis. Sheared coefficients on data for adults were used to calculate scores for the young. (Data supplied by Michael L. Smith.)

horizontal (Fig. 4.3.8b). From the alignment of the clusters of adult *robusta* and the juveniles represented by round symbols, we infer that they are expressing the same size allometries, and that they represent the same species. Coefficients are shown in Table 4.3.3.

4.3.5 Shearing in the presence of secondary factors

Consider a data set of distances measured upon specimens of two or more taxa that vary not only in general size but also in some significant subgrouping characteristic or "secondary factor," such as sex or habitat, that interacts with both taxon and size in determining shape.

If the secondary factor is confounded with size—all the males large and all the females small, for instance—then the factor S of "general size" that we compute from the pooled within-taxon covariance matrix $\mathbf{Q}'$ of log distances, unfortunately, expresses that secondary structure within taxa. Mean shape difference between the sexes at constant general size would be confounded with mean size difference, as in Figure 2.2.7a. The "size factor" S extracted from this matrix $\mathbf{Q}'$ includes aspects of sexual dimorphism at constant size. That is, it is partly a sex factor, so that subsequent inferences about shape difference factors like H are generally biased.

Such a problem calls for a modification of the shear technique to acknowledge the secondary factors. The problem does not arise if there is no difference in shape at constant size for different values of the secondary factor, as when one sex differs from the other by hypermorphosis. There is another situation in which the existence of a

4.3 The Shear: Size-Free Discrimination

subgrouping makes no difference for the interpretation of the sheared shape factor H: whenever all the taxa have the same frequency distribution upon the secondary factor, e.g., the same sex ratio. In such a case, the inappropriately pooled regressions underlying S and H have approximately the same bias—for example, the same effect of sex on shape. The resulting shape differences at constant "size" are nearly unaffected by the conflation of sex effects with size effects. This phenomenon is analogous to the fact that entering an additional predictor into a multiple regression does not alter any coefficients previously computed if the new predictor is uncorrelated with all those previously entered.

When the secondary factor affects shape at constant size within taxa and the taxa differ in their distributions upon the secondary factor, it is necessary to adjust the within-group covariance matrix $\mathbf{Q}'$ for the secondary factor. When the secondary factor is a discrete categorization, like sex, we use the matrix $\mathbf{Q}'$ of covariances computed *within category* within group, e.g., the pooled sex-specific within-taxon covariances. In other circumstances, the secondary factor is continuous (such as temperature) or is itself a factor (such as habitat) based on several indicators. In these cases $\mathbf{Q}'$ becomes the covariance matrix of the *residuals* of the log distances after their dependency on the secondary factor, or its indicators, has been explicitly regressed out. Underlying either of these maneuvers is the assumption that "general size" has approximately the same meaning for different categories (or scores) of the secondary factor, just as it must for the different taxa.

Once the size factor S is computed as the first principal component of this modified $\mathbf{Q}'$, we shear it out of the complete covariance matrix $\mathbf{Q}$ as before. The shape information that remains may show identifiable effects either for taxon or for the secondary factor. Interpretation of differences in a single distance at constant size, free of secondary factors, would proceed as an analysis of covariance by taxon, using the secondary factor as a covariate. Analysis of sheared factors like H divides them into two parts: one for the effect of the secondary factor on shape, and one for the effect of taxon. Let H_{pred} be the predicted value of H from regression upon the secondary factor or its indicators, with H_{res} the residual of H after that regression. Then the loadings for the effect of the secondary factor upon shape are the covariances of H_{pred} with the measured distances, one by one, and the loadings for the effect of taxon are the covariances of H_{res} with the same measured distances.

An alternate computation of these same loadings for the effect of taxon upon shape identifies H with the factor remaining after *both* S and the secondary factor are regressed out of the pooled matrix $\mathbf{Q}$ variable by variable. That is, we might reserve the term "shape difference factor" for the variable H_{res} of the preceding exposition. The choice between these is a matter of whether the secondary factor is fundamentally an adjustment, such as the pelvic hinge factor of Figure 4.2.10, or instead a biological cause commensurate with taxon.

When the secondary factor has no effect on shape, the quantity H_{pred} has no variance, and the computation reduces to the case of no secondary factors (Fig. 4.2.6). When, instead, the taxa under study differ considerably in their distributions on the secondary factor (sex, habitat), the loadings on H_{res} that result may differ significantly from those computed using an inappropriately pooled "within-group" covariance matrix.

4.4 Computing Average Forms from Truss Networks

Sheared principal components are useful for quantifying and describing shape differences among populations in a way that is robust against within-group size effects, i.e., for partitioning the effects of size and shape. In the ensuing analysis the forms still vary in size; in fact, one of the axes that results represents a size-measure. If we instead wish to control for size, to derive "average" shapes by standardizing the effects of size and allometry, a different approach is needed. In Section 3.2 we showed how the truss network of distance measures on the form can be used to map landmarks by "relaxing" nonplanarities. We may extend the method to produce single representative forms for subsequent descriptive analyses by averaging within-sample size variation measure by measure, and then reconstructing and mapping the form implied by the set of average distances.

There are five steps to this procedure.

1. Choose landmarks on the body outline and connect them appropriately with line segments to form a truss network.

2. Measure these distances on a sample of individuals (and check them by individual truss reconstructions).

3. Diagnose the effects of body size and allometry on the measured truss characters by log-linear regressions of the measured distances upon some composite measure of body size.

4. Choose a **standard body size** and compute the predicted values of all the distances, as estimated by the log-linear regression functions at that size.

5. Map the coordinates of the landmarks by using the predicted distances in the mapping procedure described in Section 3.2. When the standard size is a population average, we have produced an averaged form by reconstruction from the average measurements.

The composite size measure used in steps 3 and 4 could be the first within-group principal component of the covariance matrix of the log-transformed distances (Sec. 2.2.1). This is the latent (unmeasured) variable that optimally describes the joint log-linear covariation in all distance measures simultaneously. One might also use the log mean truss-element length or the mean log truss-element length as measures of size. They are approximately invariant under the operation of flattening the truss and are very highly correlated with the usual first principal component, but may be calculated for each individual without considering the rest of the population.

4.4 Computing Average Forms from Truss Networks

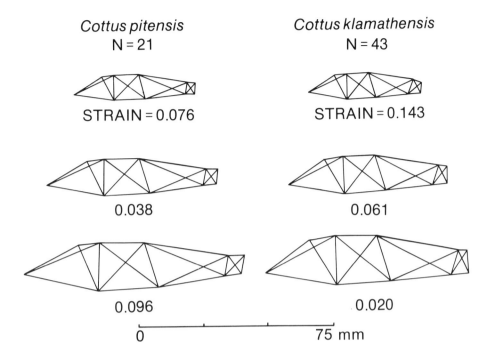

Figure 4.4.1 Averaged forms for two species of *Cottus* at three different standard body sizes. Reconstructed forms, scaled in millimeters, represent body lengths of approximately 45, 64, and 81 mm.

Partitioning of size variation by univariate regression has some disadvantages (Sec. 2.2.2.1 and Humphries et al., 1981), primarily because the resulting character adjustments are not necessarily optimal for, or even concordant with, trends of character covariation. Nevertheless, because the mapping algorithm will effectively recover the most reasonable average configuration and the strain statistics provide measures of the discord, univariate regression is appropriate in this context. The procedure results in a composite geometric form which represents the average form of a sample of individuals, standardized to some arbitrary body size, which can be drawn, examined, and compared with others.

Because the standard size chosen for the composite form is arbitrary, it may be varied to allow a direct comparison of body shapes at different sizes. Illustrated in Figure 4.4.1 are averaged forms for two closely related species of western North American sculpins, *Cottus pitensis* and *Cottus klamathensis*, at three different composite body sizes scaled in terms of the original unit of measurement (millimeters). A change in standard size alters all distance measures simultaneously, each at its own allometric rate. By "growing" the form of each species we display the composite change in form resulting from allometric influences of body size on shape.

The resulting reconstructed forms are valid only to the extent that the individually predicted distance measures describe a reasonable configuration. This becomes less

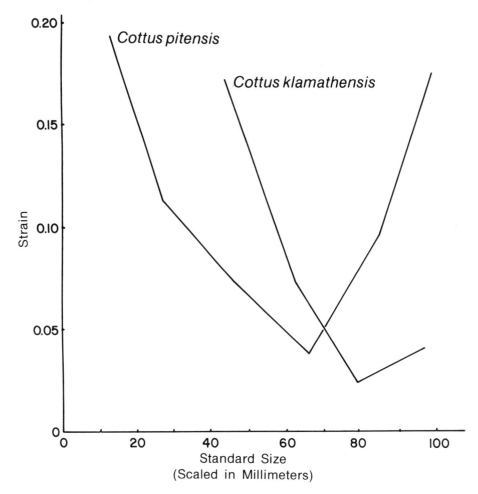

Figure 4.4.2 Dependency of strain on the size at which a truss is standardized. At extreme sizes beyond the observed range of the data, the individually predicted distance measures may fail to conform, resulting in increased strain in the relaxed truss.

likely at extreme sizes (Fig. 4.4.2). The theoretical limit to fitting distance measures is the triangle inequality, but for any empirical data set the log-linear model may fail when allometric trends are extrapolated much beyond the range of the data.

Composite averaged forms will be useful for any analysis in which standardized forms are required for comparison. They are particularly conducive to biorthogonal analyses of shape difference. In Section 4.6.2 we will apply averaged forms to the study of such shape differences within and among species. The role of the size factor may then be played by any factor that needs to be controlled, such as the positional factors of Section 4.2.

4.5 The Method of Biorthogonal Grids

4.5.1 Introduction

The methods we have discussed for analyzing shape-change are based upon a sampling of linear distances measured between landmarks over the forms. Such analyses omit several crucial sources of information about data and comparisons. Use of measured lengths restricts us to modeling forms as polygons, ignoring important differences of boundary curvatures. Unless the forms are archived as a system of trusses (see Sec. 3.2), some important data may be omitted or underrepresented because the measurement scheme does not ensure even areal coverage. All the multivariate results are highly dependent upon the particular set of distances measured, merely a sample from the infinite range of possibilities (see Fig. 3.1.2c), and may have no access to the extremal comparisons. If the orientation of our distance measures does not correspond, by accident or design, to the principal directions of shape-change (see Sec. 2.1), then our description of the differences between forms will be inadequate. This failing is quite common inasmuch as distances are generally taken in alignment with features of the form (axes of symmetry or greatest diameters, for instance) and totally ignore the matter of measuring shape-change to which they will unfortunately be applied. The report of change along a given measured length, such as a loading, is the weighted average of changes over the element of length traversed by the measurement. Inhomogeneities, such as shrinkage in the head region together with elongation posteriorly, are lost to measures such as "predorsal length." The measurements must be taken separately, region by region.

The method of biorthogonal grids was designed as a response to all of these sampling problems. It allows us to find and quantify the principal directions of shape-change, locally and globally, and is well suited for curvilinear data, such as the boundaries of real biological forms. The method ensures evenness of areal coverage and inclusion of extremal comparisons.

4.5.2 The biorthogonal method

4.5.2.1 The deformation.—The biorthogonal method models shape-change as a deformation, a single mathematical object that transforms the entirety of one form into another in accordance with biological homology. Such modeling or distortion requires smooth correspondence between forms without tears or folds and without sudden appearance of tissues. Landmarks on the boundary may not move to the interior of the form, nor conversely. Although these assumptions render grid distortions unsuitable for some transformations (e.g., if a fin is present in only one form), most comparisons of interest to systematists will not overtly violate these assumptions.

How shall we display this sort of model, which discusses two forms in a single mathematical object? We build upon the method of Cartesian grid transformations (Sec. 2.1) introduced by D'Arcy Wentworth Thompson in 1917. Thompson's method begins with a Cartesian coordinate system placed over one form, then distorts the grid into a curvilinear grid upon the second form. The two best known examples are the transformation from *Diodon* to *Mola*, and the transformation of a "human" skull to those of a chimpanzee and a baboon (Figs. 4.5.1 and 4.5.2). Unfortunately, Thompson left his readers no rules by which construction of these grids could be standardized. As a result, Thompsonian grids have traditionally been drawn in an ad hoc manner. The lines of the transformed grid often curve capriciously or extend beyond the form in the absence of any data. A few recent authors have employed grids of exactly this style (Gans, 1960; Rosen and Bailey, 1963; Tattersall, 1973; Potter and Sweet, 1981; Liem and Kaufman, 1984; other examples are cited in Gould, 1966, or Bookstein, 1978a).

A response to this problem was developed independently by Tobler and by Bookstein, both in 1977. The transformation may be successfully standardized in the form of an interpolation between landmarks, which maps one form homologously onto another, i.e., in exact consistency with data at any level of detail, and is mathematically smooth throughout the regions where there is no data (Bookstein, 1978a; Tobler, 1978).

4.5.2.2 The biorthogonal coordinate system.—Serious difficulties remain if these maps are to be usefully compared among themselves, for they are still not amenable to quantification. The deformation has been applied to an arbitrary Cartesian coordinate system not fitted to the change we wish to describe. It is highly regular, but the grid into which it is distorted has no mathematical regularities at all. Lines straight in the one form become curved in the other, and intersections that were perpendicular and symmetrical in one form are asymmetrical in the other. The appearance of curvature and of "shear" where there were none before are forced by the method and therefore are not particularly meaningful. General enlargement of the grid spacing from region to region can be detected, and also general rotation of the grid lines from the vertical/horizontal orientation with which they began. But the details of the transformation, curve by curve—the bending-apart, curvature, and angulation of the separate grid lines—are quite impossible to quantify meaningfully as separate features.

Sneath (1967) attempted a regularization of the transform as a pair of **trend surfaces**, one for each of the original Cartesian coordinates. He expected that information about the transform would be inherent in the separate regression coefficients. This is partly true—particular shapes of grid are, to some extent, legible in trend-surface form; but the descriptions of transformations are not invertible. It is not possible to describe the transformation of a "general," curvilinear grid into a Cartesian grid. In any Cartesian-based computation, whether in Thompson's style or Sneath's, it makes a great deal of difference which form you start with. Transforming *Mola* (square grid) to *Diodon* (curvilinear grid) gives a visual distortion (or set of trend-surface regression coefficients) which is not the inverse of that obtained for the original transformation in any simple way.

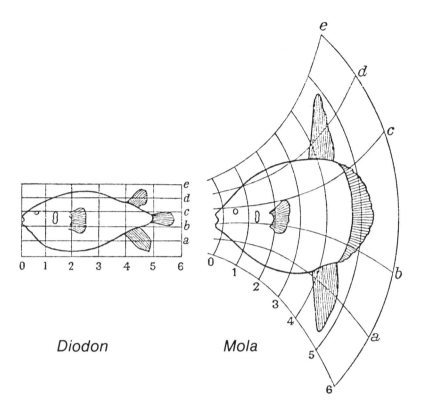

Figure 4.5.1 Cartesian deformation of *Diodon* into *Mola*. (From Thompson, 1961: figs. 154, 155.)

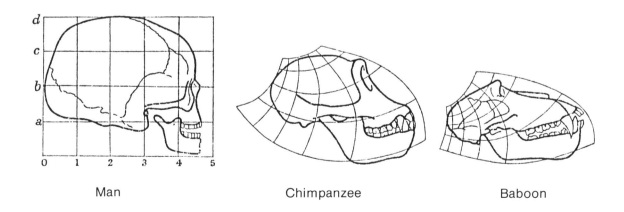

Figure 4.5.2 Cartesian deformation of primate skulls. Deformation of a "human" skull to that of a chimpanzee and that of a baboon. (From Thompson, 1961: figs. 177, 179, 180.)

The solution to this unfortunate dependence of the method on the coordinate system one starts with is the use of the local principal directions of the transformation (see Sec. 2.1), the unique pair of line-elements that is at 90° both before and after the transformation. The application of this notion in morphometrics was first hinted at in Richards and Kavanagh (1943, 1945), but its routine application, which requires high-speed computers, was delayed until the late 1970's.

For computation of biorthogonal grids, there is first a standardization: a mesh of points is placed over one form and the interpolation given by homologous data is used to map all these points smoothly into the picture of the second form (Fig. 4.5.3a). At the **mesh points**, a subset of the infinity of intersections in the coordinate system, the pairs of biorthogonal axes may be conveniently displayed (Fig. 4.5.3b). These biorthogonal axes are oriented along the directions of maximum or minimum local rate of change of length. You will recall that this relative stretch or shrink of the axes, termed the dilatation, is not the ratio of axis lengths, perpendicular elements, in one form, but rather is a measure of the change in length of each individual axis from one form to the other. Now that the transformation is characterized unambiguously as a change of lengths within one coordinate system, it is automatically invertible (by inverting the dilatations), and the choice of the first form becomes moot.

It is a simple conceptual step to describe transformations of larger shapes: we find the set of curving lines that connect the biorthogonal axes of the local areas. In practice the method has four steps, corresponding to the frames of Figure 4.5.3a–d: (1) a square mesh is placed within one form and interpolated for the second form; (2) biorthogonal axes are then computed at each homologous pair of mesh points by an analog to the procedure of Appendix A.2; (3) a set of curving lines is then derived from the principal directions of the axes by integration; and (4) dilatations are displayed along these curves ad lib. Details of these steps may be studied in earlier publications (Bookstein, 1978a, 1978b).

Up to selection of particular grid lines, we prescribe a coordinate system completely by the biorthogonal property. We have done this so as to describe the transformation completely by way of its dilatations, computed along the curves chosen by the researcher. The curves and dilatations, which sample the same information as does the mesh of crosses, describe the change in shape of the outline by a **tensor field**: a deformation of the interior by particular rates in particular directions (see Sec. 2.1). The subject of the analysis is not the pattern of curves on the forms separately, but rather the transformation between the forms as expressed in the dilatations at comparable positions in the grid. This is in no way an artifact of the engridment of the first form relative to its own accidental coordinate system. Indeed, the forms may be oriented in any manner. As for triangles (Sec. 2.1), these directions involve both forms for their computation, and cannot be determined by consideration of either form separately.

The flaw in D'Arcy Thompson's original method of construction is the lack of symmetry: a square grid for one form, but for the other a grid having no special properties at all. In the biorthogonal method, the appropriate grids for a pair of shapes

4.5 The Method of Biorthogonal Grids

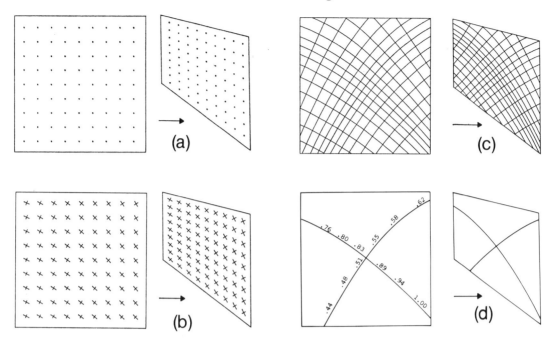

Figure 4.5.3 Biorthogonal analysis of the transformation of a square into a quadrilateral. (a) The smooth interpolation of mesh points within the homologous outlines results in a one-to-one mapping between the forms. (b) The directions of maximum and minimum rate of change (the biorthogonal directions) are computed. These represent a sampling of the tensor field relating the forms. (c) These principal directions are integrated to form curves which represent a sampling of the biorthogonal grid. (d) The ratio of homologous cross arms is the value of the dilatation at each point of the mesh. (After Bookstein, 1978a: fig. VI-6.)

have the same formal property, that all grid intersections are at 90°. This single criterion of biorthogonality essentially determines both grids at once. In this way we have reduced all change of shape to gradients of differential directional change or growth, without shear. Using this grid as a coordinate system for the change, we can measure shape change without measuring shape at all. The technique explicitly reveals the regions, directions and gradients that are worth investigating.

4.5.3 Examples

Biorthogonal analysis, as presented here, deals with change in two dimensions,† deformation in the plane of the digitized outline. It is thus desirable to choose

---

† Although the theorems underlying the biorthogonal method are valid in three-dimensional space, the computer programs have not been written. Cheverud et al. (1983) suggest a finite-element procedure for the first step, the deformation.

landmarks that lie in one plane (e.g., a flat bone or a midsagittal plane) or that may be projected onto one plane without much loss of information.

The abstractions of form used for biorthogonal comparisons comprise a selection of landmarks connected by straight line segments or curves. There is no general rule for choosing the number of landmarks, except that there must be more than two and that they should ensure representation of shape- and size-change over the entire form. Boundary curvatures may be modeled as arcs of circles or other conic sections, by constraining them to pass through additional digitized points which are not homologous from form to form. These **helping points** help determine the **analytic boundary** but not the **boundary homology function** along it (see Sec. 3.3.2). Recall that a helping point is a point with only one meaningful attribute—location (but not homology) normal to the line through its neighbors. In other words, it is measured with only one coordinate.

The choices of landmarks, inclusion of boundary curvature information, and regions for omission from analysis are all dependent on the research context. It may be possible to omit landmarks or regions or to straighten boundary arcs into segments without sacrificing biologically important information. For instance, one might do morphometrics at different geometrical scales by analyzing the component parts of a form (e.g., fin, head) separately.

4.5.3.1 Tetraodontiform fishes.—Of the transformations D'Arcy Thompson drew, the most famous is that relating *Diodon* and *Mola* (Fig. 4.5.1), two genera of tetraodontiform fishes. Thompson used simplistic outlines of *Diodon* and *Mola*; our data are derived from detailed skeletal figures in lateral view (Tyler, 1980). For these analyses we use nine boundary landmarks (Fig. 4.5.4) joined into an outline by five straight segments and four shallow circular arcs, each determined by one additional helping point. For convenience of graphical display, the outlines are scaled to the same length from tip of snout to origin of dorsal fin.

The four steps of the biorthogonal analysis (see Sec. 4.5.2) are illustrated in Figs. 4.5.5a–d. Transformation of the Cartesian grid placed over *Diodon* (Fig. 4.5.6a) results in a curving grid over *Mola*. This step, from Cartesian to curving grid, closely resembles the interpolation that Thompson drew by hand (Fig. 4.5.1). Were lines drawn to connect the mesh points in *Mola* according to their row and column in *Diodon*, we would roughly approximate Thompson's original (curving) grid, formed of pieces of "hyperbolas" and arcs of "concentric circles" centered on a point somewhere beyond the snout of the fish. Unlike Thompson's method, however, the interpolation of the Cartesian grid, or set of mesh points, is used only to approximate, not to characterize, the smooth deformation. Rather, it is the computation of the biorthogonal axes at each of the mesh points that characterizes the transformation (Fig. 4.5.5b). Sampling these principal directions of change along curves covering the form (Fig. 4.5.5c), then displaying the values of the dilatations along these directions (Fig. 4.5.5d), we complete the steps of the biorthogonal analysis.

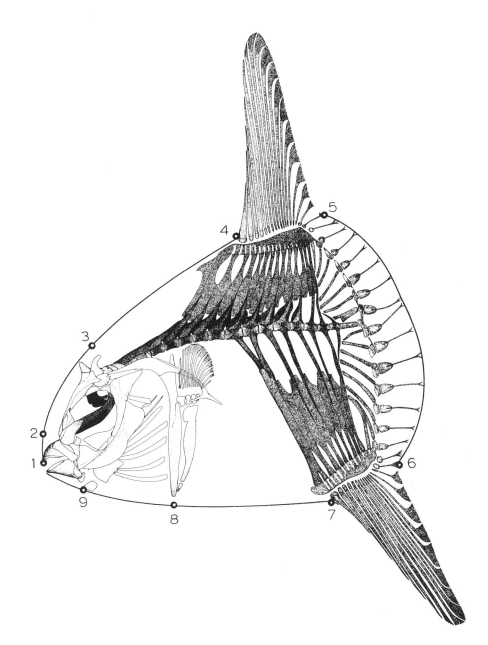

Figure 4.5.4 Construction of an outline for biorthogonal analysis of tetraodontiform fishes. The nine homologous boundary landmarks are, clockwise: (1) tip of snout, (2) point on outline nearest to posterodorsal tip of premaxilla, (3) point on outline nearest to basicranial base, (4) dorsal fin origin, (5) junction of dorsal and caudal fins, (6) junction of caudal and anal fins, (7) anal fin origin, (8) ventral tip of cleithrum, (9) point on outline nearest to posterior tip of the angular bone. These points are joined together by five straight line segments and four shallow circular arcs. (Modified from Tyler, 1980: fig. 306.)

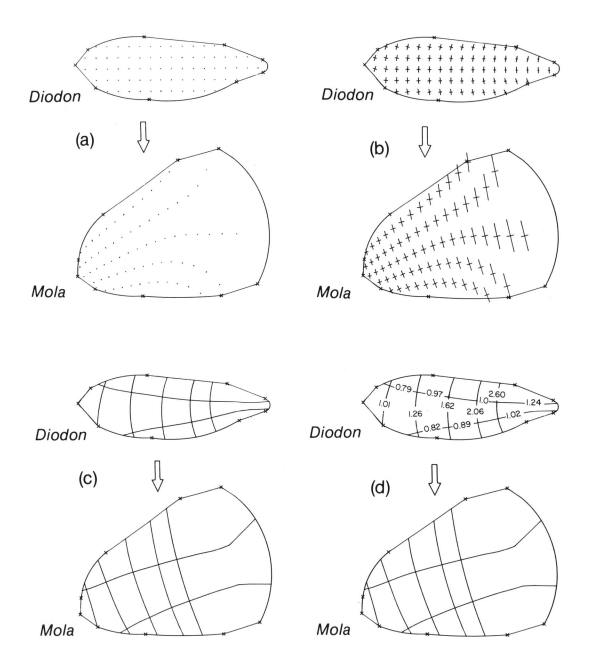

Figure 4.5.5 Biorthogonal analysis of the transformation of *Diodon* to *Mola*. (a–d) The four steps of the analysis as described in Figure 4.5.3 and in the text.

4.5 The Method of Biorthogonal Grids

In this example, the principal directions of change are oriented along the anteroposterior and dorsoventral axes of *Mola* (Fig. 4.5.5c). That the grid is square in *Mola* is a finding of the analysis, not an artifact of the starting coordinate system; the homologous principal directions of change are not squared in *Diodon*. Instead, the anteroposterior curves converge on the caudal peduncle and the dorsoventral curves are convex toward the head region. Along the anteroposterior curves are distinct gradients—from relative shrinkage (0.78) in the head region to stretch (1.24) in the caudal peduncle. The dorsoventral curves show little dorsoventral gradient but reveal an even stronger gradient anteroposteriorly. The curves near the head bear dilatations of 1.2, while one near the caudal region has an average dilatation of about 4.5. The transformation from *Diodon* to *Mola* might be thought of as a squaring of *Diodon*'s grid, involving both stretch and shrink.

The very large posterior expansion of *Mola* relative to *Diodon* is almost at the limits of the deformation model; its consequences for the analysis are instructive here. In Figure 4.5.5b, note that there are only two columns, with two mesh points each, in the relatively narrow caudal peduncle of *Diodon*. The expansion of this region in *Mola* has too few mesh points and biorthogonal crosses to fill the large convexity of the hypural complex. The most posterior dorsoventral curve should have been distorted by the large displacement in the posterior boundary arc, but was not. If we increase the number of mesh points posteriorly in both forms, we make the analysis more sensitive to this change in boundary. In this separate analysis of the posterior portions of the forms (Figs. 4.5.6a–d), the boundary effect can be seen clearly. The effect does not propagate very far into the interior of *Mola*. The preceding whole-body analysis is not misleading, merely insufficiently detailed.

What if we were to invert the comparison, *Mola* to *Diodon* (Fig. 4.5.7)? The interpolation pattern appears different (Fig. 4.5.7a vs. Fig. 4.5.6a) only because the transformation has been replaced by its inverse. Nevertheless, the biorthogonal grids are nearly identical, up to choice of individual grid lines, to those of Figure 4.5.6c. The dilatations (Fig. 4.5.7d) are approximate reciprocals of those in the original transformation and reveal the same gradients of change. No matter which form bears the Cartesian grid, the biorthogonal method yields the same depiction of these gradients.

We may examine biorthogonal deformations among members of the Molidae from an evolutionary perspective. These tetraodontiform fishes are hypothesized (Tyler, 1980) to have shared their most recent common ancestor with the lineage giving rise to tetraodontids and diodontids. Only three genera are known, each monotypic: *Ranzania*, *Mola*, and *Masturus*. *Ranzania* is the primitive member of this monophyletic triad (Fig. 4.5.8). The cladogenic event giving rise to *Ranzania* and the ancestor of *Mola* and *Masturus* should be expressed in the transformations from *Ranzania* to either of the derived taxa. Differences between the transformations from *Ranzania* to *Mola* and to *Masturus* may indicate the divergence of each of the latter from their common ancestor.

The transformation from *Ranzania* to *Masturus* results in a grid on *Ranzania* (Fig. 4.5.9) whose anteroposterior curves converge posteriorly except over the hypural

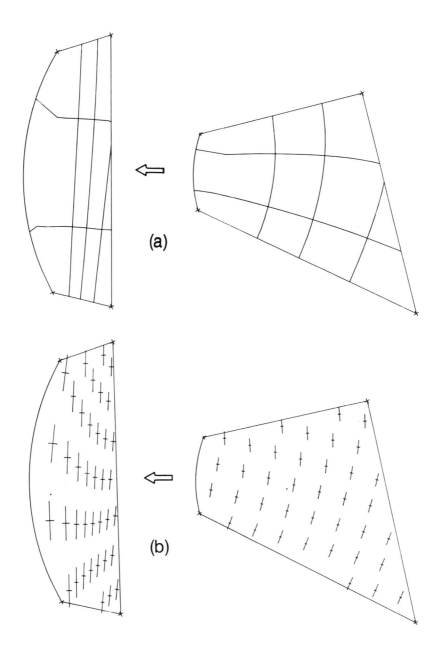

Figure 4.5.6 Biorthogonal analysis of the caudal regions of *Diodon* and *Mola*. The boundary landmarks are the dorsal fin origin, the junction of the dorsal and caudal fins, the junction of the caudal and anal fins, and the anal fin origin. The landmarks are joined together by a shallow circular arc posteriorly and straight-line segments elsewhere. Note that the straight boundary through the body of the fish is arbitrary. (a) Outlines showing the biorthogonal grids. (b) A sampling of the tensor field within the forms.

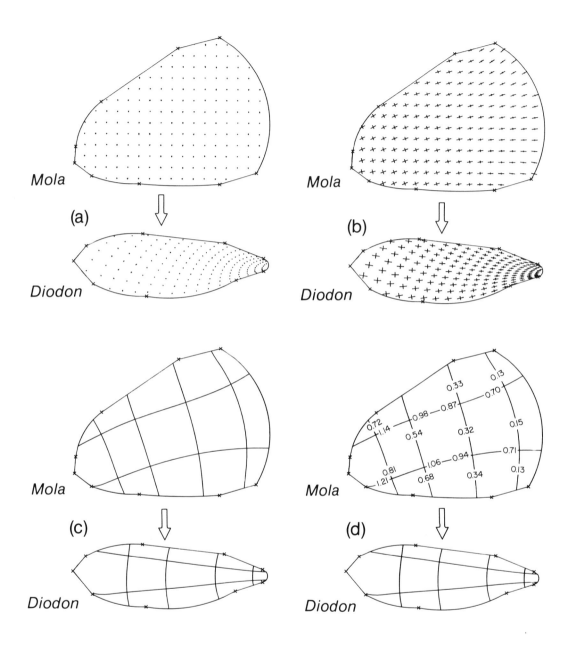

Figure 4.5.7 Biorthogonal analysis of the transformation of *Mola* into *Diodon*. (a–d) The four steps of the biorthogonal analysis as described in Figure 4.5.3 and in the text. Compare Figure 4.5.5.

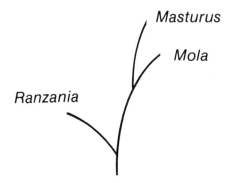

Figure 4.5.8 Probable phylogenetic relationships among selected genera of tetraodontid fishes. (Based on Tyler, 1980.)

complex. Along these directions, the dilatations less than 0.95 indicate that the anterior two-thirds of *Masturus* is shrunk relative to *Ranzania*, but that beyond the origins of the median fins it is expanded. The dorsoventral grid lines lie almost perpendicular to the anteroposterior axis of the fish; their dilatations indicate the marked deepening of *Masturus*, especially posteriorly. The principal directions of deformation form nearly a square grid on *Masturus*. This overall squaring is not unlike the transformation from *Diodon* to *Mola*.

The transformation from *Ranzania* to *Mola* (Fig. 4.5.10) appears very much like that to *Masturus*: a slightly curving grid on *Ranzania*, a square grid on *Mola*. Again, the anteroposterior aspect of *Mola*, anterior to the origins of the median fins, is compressed relative to *Ranzania*, and the dorsoventral aspect is greatly expanded.

The transformations to *Masturus* or *Mola* from their primitive sister species, *Ranzania*, have in common a square grid coupled with anterior longitudinal compression and dorsoventral expansion. We infer that this commonality of deformation represents the reorganization of form (boundary and landmarks) in the lineage to *Masturus* and *Mola* after its divergence from the line to *Ranzania*. The difference in transformations (e.g., magnitudes of the dilatations) indicates the differences between *Mola* and *Masturus* relative to their most recent common ancestor.

The transformations among the Molidae are smoothly graded over the forms. In other comparisons the computed changes may vary sharply between adjacent regions.

4.5.3.2 Poeciliid fishes.—Outlines of dorsal views of skulls of poeciliids, *Gambusia affinis*, *G. beebei*, *G. manni*, and *Belonesox belizanus*, were digitized from Rosen and Bailey (1963:81). They presented certain transformations among four forms representing these taxa in the Cartesian style of Thompson (Fig. 4.5.11). The true scales for their drawings are unknown.

Twelve landmarks were chosen and four helping points were included to help define the arcs of the outline (Fig. 4.5.12). In order to avoid redundancy and limit the

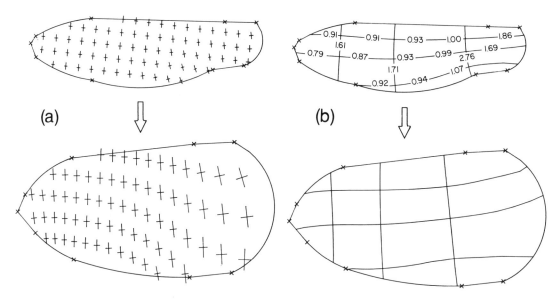

Figure 4.5.9 Biorthogonal analysis of the transformation of *Ranzania* to *Masturus*. (a) The biorthogonal crosses. (b) Selected curves and dilatations.

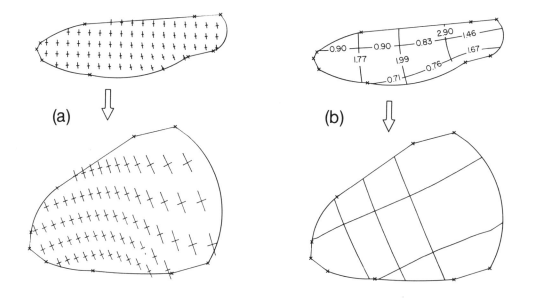

Figure 4.5.10 Biorthogonal analysis of the tranformation of *Ranzania* to *Mola*. (a) The biorthogonal crosses. (b) Selected curves and dilatations.

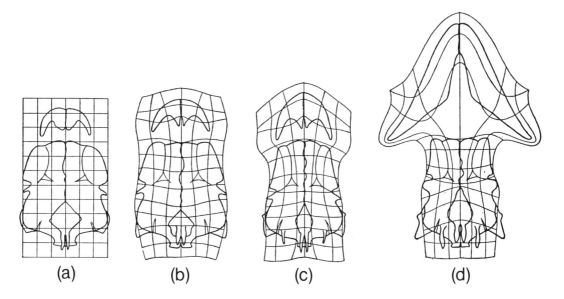

Figure 4.5.11 Cartesian deformation of poeciliids. *Gambusia affinis* (a) deformed into (b) *G. manni*, (c) *G. beebei*, and (d) *Belonesox belizanus*. (From Rosen and Bailey, 1963: fig. 35, scale omitted.)

number of internal landmarks (see Sec. 5.1), each taxon was represented by the right half of its neurocranium and upper jaw.

In one aspect the shapes here are somewhat arbitrary. In life the area between the premaxilla and neurocranium is occupied by both hard and soft tissues, in diverse planes and orientations, which are mutually displaced as the fish opens its jaws. In the absence of more precise figures, to model changes in this functional unit as deformation we join premaxilla to neurocranium with two straight lines. If this data set were to be studied other than by way of example, the assumptions and data would have to be addressed more rigorously, the homologies of the landmarks scrutinized more carefully, and the placement of the helping points better determined.

When these data were recorded, the outline for *Gambusia affinis* was inadvertently digitized twice. We thus provided ourselves with a test of the implications of digitization noise for biorthogonal analysis. Comparison of this outline to itself (Fig. 4.5.13) shows that careful digitizing introduces little noise into the data. The random spinning of the crosses and their isometry (dilatations all $1.00 \pm .01$) is an expression of the near-identity of the comparison.

In a first set of biorthogonal analyses (Fig. 4.5.14) the outlines were scaled just as in Rosen and Bailey. As the neurocrania are set to similar size, comparisons among the three *Gambusia* species reveal slight differences, primarily in the region between the premaxilla and the neurocranium. Here the straightness of the connection causes abrupt shifts in the angle of the outline. Because this line segment and its angle with the rest of the outline are arbitrary, we should not interpret their effect on the grids;

4.5 The Method of Biorthogonal Grids

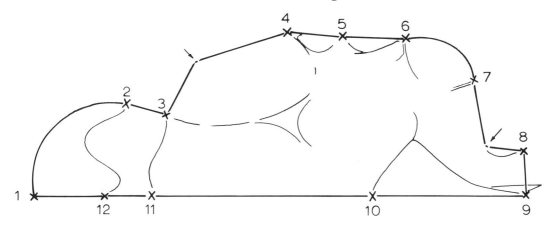

Figure 4.5.12 Construction of an outline for biorthogonal analysis of poeciliid fishes. The homologous boundary landmarks are: (1) anterior tip of the premaxilla at the midline, (2) posterolateral tip of the premaxilla, (3) junction of the supratemporal canal with the anterior border of the frontal, (4) lateral corner of the frontal at the dermosphenotic, (5) lateral extreme of the sphenotic, (6) projected intersection of the parietal, sphenotic, and pterotic, (7) posterior junction of the epiotic and the pterotic, (8) posterolateral tip of the exoccipital, (9) posterior tip of the exoccipital at the midline, (10) posterior tip of the frontals at the midline, (11) anterior tip of the frontals at the midline, (12) posterior tip of the premaxilla at the midline. Two helping points were included at the inflections of the outline at the anterolateral corners of the frontal and exoccipital (marked by arrows). The landmarks and helping points are connected by 12 straight-line segments and two shallow circular arcs. Note that the straight-line segments between the skull and premaxilla are arbitrary.

fortunately, the effect propagates inward only a short way, one or two mesh points. Other changes among the *Gambusia* species are relatively slight.

The head of *Belonesox* is distinctly larger than those of the species of *Gambusia*. When any of the three is compared to *Belonesox*, the premaxillary region exhibits dramatic change. Its anteriormost landmarks have been displaced anteriorly, its lateral tip laterally. The neurocranial comparisons are not very different from those among *Gambusia*.

A second set of analyses was performed with these skulls standardized to unit length distance from anterior tip of premaxilla to posterior tip of occipital (Fig. 4.5.15). The neurocrania of the *Gambusia* species are similar enough in size that such a standardization does not substantially alter their comparisons. The comparisons of *Gambusia* with *Belonesox* still show that the premaxillary region increases dramatically and that the neurocranium appears to be considerably reduced in size. While the dilatations are changed by a constant factor, the principal directions of shape change are unaffected by the standardization procedure.

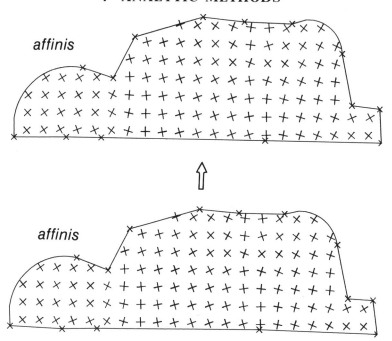

Figure 4.5.13 Biorthogonal analysis of two digitizations of the outline of *Gambusia affinis*.

The biorthogonal grids for these poeciliids differ from the Cartesian grids of Rosen and Bailey (1963) in several respects. Theirs, as usual with Cartesian grids, extend beyond the form and violate the homology map as we have computed it. More significantly, the directions of greatest and least amount of change are strikingly different. In the premaxilla of any of the *Gambusia-Belonesox* comparisons, the biorthogonal grids indicate a simple shift of the lateroposterior landmark. However, the Cartesian grids curve laterally and move out of the premaxilla completely. In the comparison among the gambusiins, the skull region in the Cartesian grids shows lines that remain oriented very nearly parallel to the original square grid placed over *G. affinis*. (There is some slight suggestion of compression or tension by convergence or divergence of the lines over certain areas.) In the biorthogonal analysis, however, the directions of change are usually oriented at 45° to the midline of the neurocranium, expressing a shearing of skull outline.

4.6 Allometric Growth and Shape-Change

The theme of this section is the discovery and description of form-change among populations each represented by specimens of varying sizes. The analyses consider both allometric growth patterns within populations and their consequences for descriptions of shape differences observed between populations. We assume that

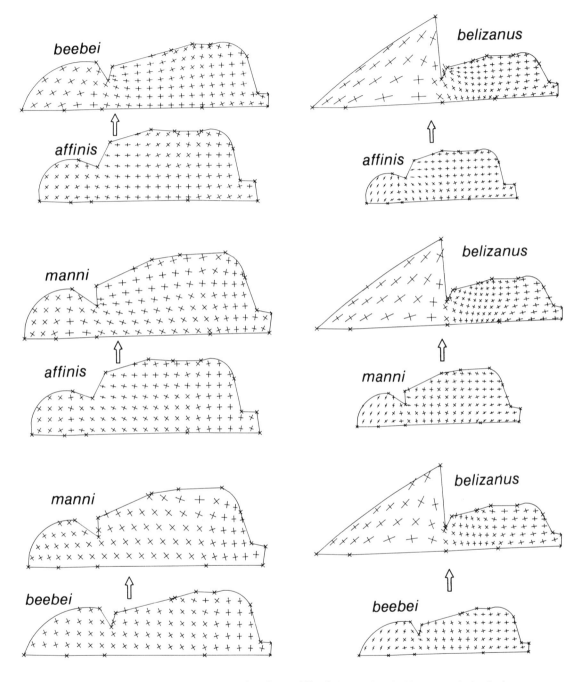

Figure 4.5.14 Biorthogonal analysis of poeciliid fishes. The skulls are scaled relative to each other as in Figure 4.5.12. True scale is unknown.

144   **4 ANALYTIC METHODS**

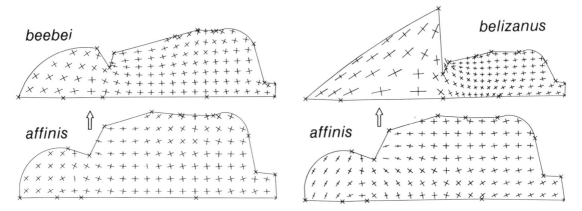

Figure 4.5.15 Size-standardized biorthogonal analysis for selected pairs of poeciliid fishes. The skulls are standardized to the length along the midline from the anterior tip of the premaxilla to the posterior tip of the exoccipital.

allometric growth can be described from a mixed series in which individuals of different sizes at the same time serve as a proxy series for a single, "typical" individual growing over time. The phenomenon of allometry will affect our discriminations whenever the distance measures that contribute to differences between species have different allometries with respect to within-group size or when the principal directions of shape-change between groups fail to align with those calculated from size series within groups. In this section, we will compare the effectiveness of biorthogonal analysis with that of factor analysis in the resolution of these interactions.

4.6.1 Shape-change of two atherinid fishes

The silverside fishes of Central America are not well understood. The current confusion regarding their systematics owes partly to the insufficiency of conventional ratios of distance measures for discriminating among different morphotypes. *Atherinella pachylepis* and *A.* sp. (=*Melaniris guatemalensis* of previous authors, e.g., Miller and Carr, 1974, Bussing, 1978) are a pair of closely related silversides from the Pacific coast that occur sympatrically in El Salvador and have been particularly problematic as to their species identities and distributions. In the analyses below, samples of *A.* sp. from Guatemala to Costa Rica will be compared to *A. pachylepis* from Panama.

4.6.1.1 Two versions of the data base.—The correspondence between factor analysis and biorthogonal analysis reflects the extent to which the analyses draw upon the same geometric data. For biorthogonal analyses, the data are curvilinear forms with landmarks. From the radiographs of individuals, ten landmarks and two pseudolandmarks were digitized as indicated by the arrows in Figure 4.6.1. We did not

4.6 Allometric Growth and Shape-Change

analyze data from the fins except at their bases. The same landmarks are indicated by X's along the analytic boundary of the form shown in Figure 4.6.2, constructed from five arcs of circles, five straight lines and two sections of parabolas. This scheme was discussed at length in Section 3.3.3. Although the reconstructed form resembles a planarian, once the head, eye and fins are drawn (Fig. 4.6.3) it becomes recognizable as the form of a fish. In the biorthogonal analyses which follow, all individuals were photographically enlarged to a common standard length; the separate dilatations of the analysis express allometries with respect to this length. Relations between dilatations of the same transformation, and comparisons of gradients between transformations, may be interpreted regardless of any such scaling.

Figure 4.6.1 Radiograph of *Atherinella pachylepis*. Arrows indicate landmarks for biorthogonal analysis.

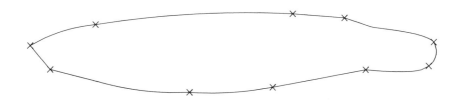

Figure 4.6.2 Analytic boundary of *Atherinella pachylepis*. The X's mark the positions of landmarks.

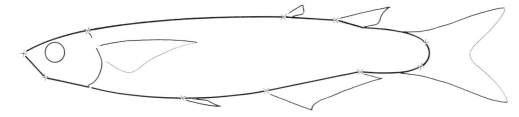

Figure 4.6.3 Analytic boundary of *Atherinella pachylepis* with head, eye and fins added.

For principal components analyses, we use 22 typical measured distances (Fig. 4.6.4) primarily oriented along the main longitudinal axis of the fish. In addition, there are two atypical measures: the diagonals from the pectoral-fin insertion to the origins of the pelvic and anal fins (Fig. 4.6.4). We will see that these oblique measures express important information about form change.

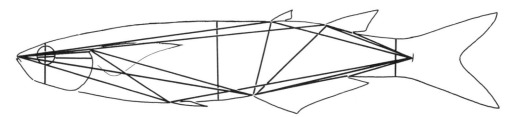

Figure 4.6.4 Distance measures for analysis of atherinids.

Our biorthogonal analysis program (Sec. 4.5) computes biorthogonal crosses at each point of a square mesh suspended over one of the forms, so providing an even sampling of rates and principal directions of form change. Reports of shape-change from principal components analysis are restricted to the measured lengths; local rates of change are necessarily averaged over the segments measured. The major differences in findings between the two styles of analysis derive from these differences in evenness of areal coverage and persistence of local information.

4.6.1.2 Shape-difference between species.—We computed principal components of the covariance matrix of the 24 log-transformed distance measures for 55 specimens of *Atherinella* sp. together with 40 specimens of *A. pachylepis*. The scatter diagram of scores on the first two principal components shows species clusters distinct in this plane and oblique to the axes (Fig. 4.6.5). To better estimate the shape-difference between these species, we removed the confounding effect of pooled intraspecific size by shearing the second principal component (see Sec. 4.3). Shearing clarifies the ordination, but preserves the tipping of the intraspecific principal axes towards each other (Fig. 4.6.6). It appears that as individuals of these species grow larger, they approach a similar configuration. This point will be taken up in Section 4.6.1.5.

The size-free shape differences between *A.* sp. and *A. pachylepis* are indicated by the loadings of the sheared second component (Fig. 4.6.7). *A. pachylepis* has a larger pectoral to anal-fin length and a longer and deeper caudal peduncle than *A.* sp.; *A.* sp. has a longer head and snout, and wider bases of the anal and second dorsal fins, than has *A. pachylepis*.

Biorthogonal analysis was performed using the pair of specimens closest to the centroids of their clusters. The distortion of the grid between the forms (Fig. 4.6.8) represents the transformation from *A. pachylepis* to *A.* sp. The anterior region of the head, especially the snout, is relatively longer in *A.* sp. Posterior to the snout, there are crossing diagonals of relative shrinkage. However, the region above the base of the

4.6 Allometric Growth and Shape-Change

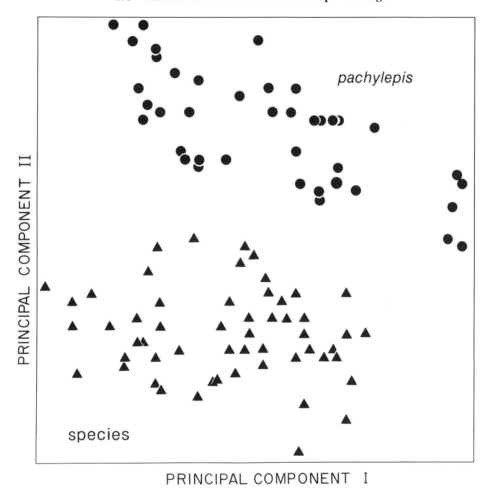

Figure 4.6.5 Scatter of scores on first two principal components for *Atherinella*. *A. pachylepis*, circles, $n = 40$; *A.* sp., triangles, $n = 55$.

anal fin is relatively expanded in *A.* sp. The posteroventrally directed curves in the anterior portion of the body and the posterodorsally oriented curves in the posterior portion of the body are directions of greater relative shrinkage than the curves perpendicular to them.

The principal components analysis corresponds well with the biorthogonal analysis. Each detects the elongation of the snout, the expansion of the region above the anal-fin base, and the oblique foreshortening anterior and posterior to the anal fin in *A.* sp. Head length has a moderately large coefficient (0.22) on the shape discriminator, indicating a larger head in *A.* sp. (Fig. 4.6.7). This loading is partially confounded with that of 0.51 for snout length, a measure geometrically overlapping with head length. The biorthogonal analysis shows that even though the snout of *A.* sp. is elongated, the head is relatively shorter postorbitally. The change in relative head

148   4 ANALYTIC METHODS

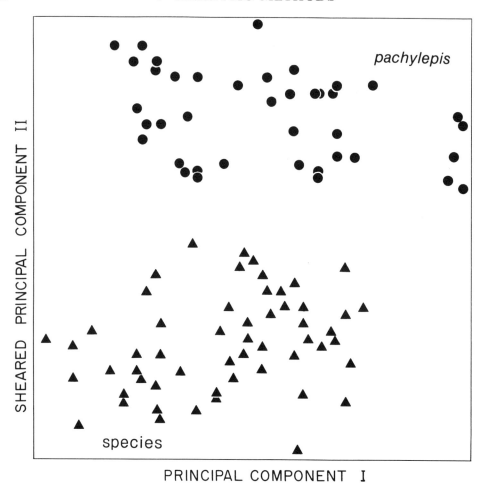

Figure 4.6.6 Scatter of scores on shape factor and first principal component for two species of *Atherinella*. *A. pachylepis*, circles, $n = 40$; *A.* sp., triangles, $n = 55$.

length between these species as indicated by principal components analysis is an average of local changes over the entire segment measured. If the relative shrinkage of postorbital head length of *A.* sp. had been equal in magnitude to the relative elongation of its snout, then the net loading for head length would approximate zero.

4.6.1.3 Growth of Atherinella pachylepis.—To understand fully the differences in shape between species, the changes in form that occur with growth must be explored *within* each species. To this end we inspect the biorthogonal grids for the transformation of the 56 mm form of *A. pachylepis* into the 112 mm form. Because these have been digitized to the same standard length, dilatations represent allometry with respect to standard length. Somewhat arbitrarily, we consider dilatations of $1.00 \pm .03$ to represent relative isometry; those greater than 1.03 connote positive

4.6 Allometric Growth and Shape-Change

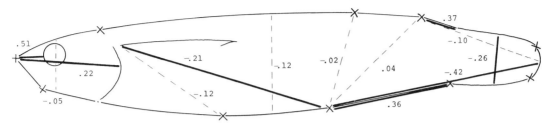

pachylepis & species

Figure 4.6.7 Discriminating distance measures for *Atherinella pachylepis* and *A. sp.* Loadings on shape factor are shown. Those distances with positive coefficients are longer in *A. pachylepis*; those with negative coefficients are longer in *A. sp.*

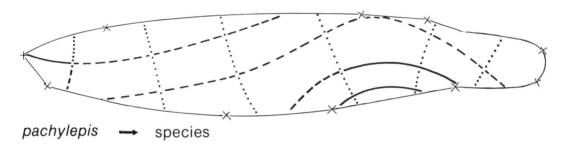

pachylepis → species

Figure 4.6.8 Biorthogonal grid distortion from *Atherinella pachylepis* to *A. sp.* A sampling of principal curves is shown. Dilatations range from 0.76 to 1.13. Those greater than 1.0 signify extension in *A. sp.* relative to *A. pachylepis*, represented by solid lines. Dilatations less than 1.0 signify extension in *A. pachylepis* relative to *A. sp.* They are coded as follows: 0.90–0.99, dashed lines; 0.76–0.89, dotted lines.

allometry, those less than 0.97, negative. In Figure 4.6.9 the curves of the grid are coded by this classification of the dilatations along them.

That figure reveals a striking finding: the principal directions of change are oriented obliquely to the anteroposterior axis of the fish rather than along the directions of typical measurements. Posterior to the head, the posteroventrally-directed curves are positively allometric, and the anteroventral curves are negatively allometric, that is, "foreshortened." We interpret this transformation as a caudal shearing of the form—caudal displacement of the pelvic and anal fins together with some size change. In other words, growth behind the head involves relative extension of the form posteroventrally. The head grows mainly in negative allometric fashion, as has been noted for other silverside fishes (Chernoff et al., 1981).

A within-group general size factor for *A. pachylepis* was estimated by the first principal component of the 24 distance measures, and its coefficients were normalized to mean square 1. The normalization procedure preserves the allometric ordering of the variables. In particular, if all variables covaried equally with size (Sec. 2.2.1), then

150   4 ANALYTIC METHODS

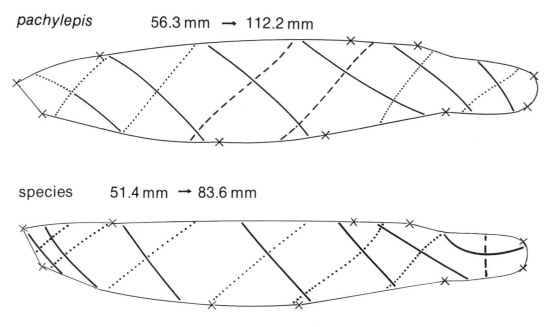

Figure 4.6.9 Biorthogonal grids for growth in *Atherinella*. *A. pachylepis*, from 56.3 mm to 112.2 mm SL; dilatations from 0.84 to 1.20. *A.* sp., from 51.4 mm to 83.6 mm SL; dilatations from 0.85 to 1.36. A sampling of principal curves is shown. The solid curves bear dilatations greater than 1.0, representing positive allometry with respect to standard length. The dashed curves bear dilatations about 1.0, representing isometry; and the dotted curves bear dilatations less than 1.0, representing negative allometry.

all the coefficients would be normalized to 1.0, indicating isometry. Coefficients greater than 1.0 indicate positive allometry with respect to general size, and those less than 1.0, negative allometry, as diagrammed in Figure 4.6.10.

Again the principal components analysis corresponds well with the biorthogonal analysis, except where the loadings on general size conceal variations of local coefficients along the measured segments. For example, in the biorthogonal analysis, along the curve drawn posteroventrally through the head the allometric coefficients increase posteroventrally (Fig. 4.6.9). Thus, the negative allometry of the head and the subsequent positive allometry are averaged into the apparent isometry of prepelvic length with respect to general size (Fig. 4.6.10). Because atypical diagonal measures have been included in the set of variables factored, the loadings from the component analysis corroborate the shearing of body form during growth observed in the grid analysis. Anterior diagonals from the upper pectoral-fin base to the pelvic-fin insertion and anal-fin origin are positively allometric with respect to posterior diagonal measurements (pelvic fin to caudal base, anal-fin origin to caudal base and caudal peduncle length). The biorthogonal analysis is clearly the more informative in this matter.

4.6 Allometric Growth and Shape-Change

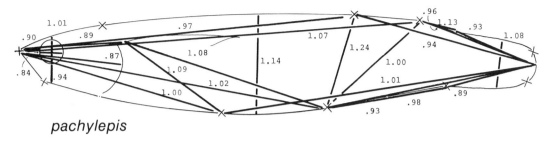

pachylepis

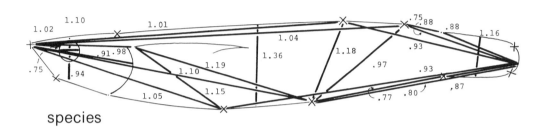

species

Figure 4.6.10 General size factors for *Atherinella pachylepis* and *A.* sp. Coefficients greater than 1.03 connote positive allometry with respect to general size; those between 0.97 and 1.03 are taken to signify relative isometry; those less than 0.97 signify relative negative allometry.

4.6.1.4 Growth of Atherinella sp.—The biorthogonal grids for the transformation from a 51.4 mm specimen of *A.* sp. to one of 83.6 mm (Fig. 4.6.9) are similar to those for growth in *A. pachylepis*. The principal axes of shape-change lie on oblique diagonals, with dilatations larger posteroventrally. But growth in *A.* sp. differs in having larger dilatations anterodorsally in the anterior part of the head; for instance, the snout grows without negative allometry with respect to other dorsoventral extensions. In the middle of the body the anteroventral curves are more negatively allometric in *A.* sp. without any gradient of dilatation. The caudal peduncle of *A.* sp. grows more positively allometrically than that of *A. pachylepis*. (Note that one of the principal directions of change is oriented along the depth of the peduncle.)

Again there is general concordance between the findings of the two styles of analysis. The isometry of snout length in the principal components analysis (Fig. 4.6.10) averages across the principal axes of stretch and shrinkage displayed by the biorthogonal analysis (Fig. 4.6.9). In both analyses, posteroventral directions are found to be positively allometric with respect to anteroventral directions.

Yet the analyses do not agree totally; there are some discrepancies. In the principal components analysis, body depth and least depth of the caudal peduncle appear to be strongly positively allometric with respect to general size, but from the biorthogonal analysis we would predict these measures to be isometric or only weakly positively allometric. Body depth bisects the angle between the principal curves and bears the average of their dilatations (Sec. 2.1); the caudal peduncle was identified as a region of isotropic change. We suggest two possible reasons for the discrepancy. (1) Body depth and caudal peduncle depth are not homologous from form to form—they are not taken between landmarks. The analyses thus begin with different collections of geometric information. (2) At present, biorthogonal grid analysis can only analyze the transformations between two forms, and the pair of specimens we chose may not be typical in regard to these measurements. This can be overcome by the generation of average forms from the truss pattern of measurements (see Sec. 4.6.3).

4.6.1.5 Explanations of shape-change.—We noted in Figure 4.6.6 that the shapes of these taxa (as discriminated by the shape-difference factor) were converging with growth. Because the grid patterns for growth in the two species are so similar, this finding is real and not an artifact of projection onto the plane of the scatter. In fact, this convergence in form can be identified in certain differences of the within-species allometries as viewed in either the factor analyses or the grids. In the factor analyses, the loadings for body and caudal peduncle depths and for the diagonals from the pectoral fin are more highly allometric in *A.* sp. than in *A. pachylepis*. The within-species biorthogonal analyses convey the same information—larger dilatations are relatively larger in *A.* sp. than *A. pachylepis*, even though the transformation within *A.* sp. is across a smaller range of standard lengths. These larger dilatations correspond to distances along which *A.* sp. began at a relative disadvantage. Except for the head regions, these closely related species change form with growth in similar fashion but at different rates, so that *A.* sp. seems to be "catching up" with *A. pachylepis*, achieving a given amount of form change over a smaller size range. Such differences in the relative rate of shape change with respect to size change will be encountered again in Section 5.3.3.

4.6.1.6 Conclusions.—Our understanding of the differences in form between these silversides is enhanced by comparison of the principal directions of form-change and their allometries. That the principal axes of growth in these taxa are not oriented along lines of traditional measurements is an important finding. Their true orientation, obliquely across the body, suggests a new non-standard system of measured diagonals crossing along the body (see Secs. 3.2 and 4.6.3-4). The results of principal components analyses conform with biorthogonal analyses to the extent that the distance measures correspond to the principal directions of shape-change. Generally, the biorthogonal analysis bears more detail of both spatial position and angular orientation. In particular, loadings near zero on long measurements may be concealing areas of differential change.

4.6 Allometric Growth and Shape-Change

4.6.2 Ecophenotypic differentiation in a cottid

The preceding example described how convergence to similar body forms in adults of two closely related species may result from a difference in patterns of growth and allometry. In contrast, divergent patterns of growth among populations of a single species existing in different habitats may lead to notably different body shapes. For example, lake and stream populations of a single fish species may differ markedly in morphometric as well as meristic characters (Hubbs, 1941). We describe such a case to illustrate the role in intraspecific divergence of differences in allometry. In doing so we will also contrast the utility of several methods of analysis for describing these effects.

4.6.2.1 Shape differences between ecophenotypes.—Strauss (1980), in a conventional multivariate statistical analysis of geographic variation in the freshwater sculpin *Cottus cognatus*, discovered a clear difference in morphology between two samples from Lakes Cayuga and Ontario in New York and six samples collected from streams throughout Pennsylvania and Michigan (Fig. 4.6.11). The data for the principal component analysis comprise logarithms of the 22 traditional morphometric measures of Table 4.6.1. These are adjusted for body size by separately regressing each variable on the log mean distance-measure per individual, then replacing the original measurements with the regression residuals (Thorpe, 1976, 1980; see Section 4.4). If the effects of size have been adequately accounted for by this method, then scores on the first among-group principal component are size-independent and should express the primary differences in form among samples.

In this analysis lake and stream forms are clearly distinguishable (Fig. 4.6.11a) but the loadings (Table 4.6.1) are difficult to interpret. The size-adjustment did not lead to a balance of positive and negative loadings. All substantially non-zero loadings on component I are negative except for two: the distances from the pectoral-fin base to the snout and to the base of the caudal fin. It is difficult to interpret this contrast geometrically. The consistency of sign among most of the loadings might reflect a difference in "robustness" between the lake and stream forms, that is, a difference in mean size between samples of the same length. The effects of size would not have been completely removed from the data if the within-group regression slopes on mean size were not equal among samples. Such a difference might occur if, for example, the allometries themselves were dependent upon some exogenous factor varying across habitats. Nine of the 22 regression slopes were significantly different between the groups. Thus the results of the conventional analysis are ambiguous, and might preclude the discovery of any real differences in patterns of growth among ecophenotypes.

A more satisfactory analysis of shape difference between morphotypes, still with respect to this conventional data set, is provided by sheared principal components (Sec. 4.3). Because the discrimination due to difference in mean group size has been partitioned out, group separation (Fig. 4.6.11b) is not as complete as it is in the analysis of residuals. However, the loadings on the sheared second component reveal a

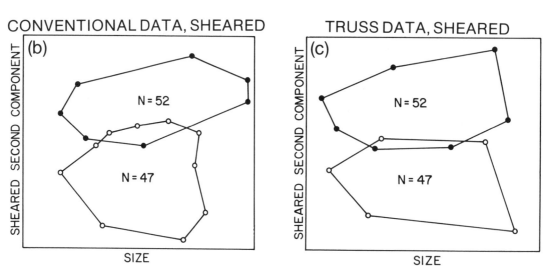

Figure 4.6.11 Principal components analyses of samples of the sculpin *Cottus cognatus* collected from lakes and streams. Lakes–closed circles; streams–open circles. (a) The first two components of the residuals of a conventional character set, formed by separately regressing each variable on a multivariate measure of body size. (b) The sheared second principal component (*H*) of the conventional character set, plotted against the first pooled within-group component (size, *S*). (c) The sheared second component (*H*) and within-group size axis (*S*) of the truss character set.

4.6 Allometric Growth and Shape-Change

Table 4.6.1 Principal component loadings of 22 conventional morphometric measures on *Cottus cognatus*.

Character	Component loadings	
	PC I of Residuals	Sheared PC II
Head length	−.03	.16
Premaxilla length	−.23	−.04
Maxilla length	−.19	−.08
Vomerine tooth-patch width	−.21	−.54
Eye diameter	.04	−.08
Postorbital head length	−.13	−.07
Least depth caudal peduncle	−.37	−.50
Maximum body depth	−.21	−.08
Body width above anus	−.24	−.08
Length first-dorsal fin to hypural	.04	.04
Length second-dorsal fin to hypural	−.21	.08
Length anal fin to hypural	−.25	.07
Length snout to pectoral base	.14	.35
Length pectoral base to hypural	.21	.35
Length first-dorsal fin base	−.02	.08
Length second-dorsal fin base	−.08	.07
Length anal fin base	−.08	.12
Length of depressed second-dorsal fin	−.35	−.08
Length of depressed anal fin	−.33	−.08
Length of pelvic fin	−.34	−.23
Standard length	.03	.17
Total length	−.25	.10

general contrast between two sets of characters. The many redundant distance measures parallel to the body axis are contrasted with vomerine width, caudal-peduncle depth, and pelvic fin length.

To better describe the overall difference between the two body forms, 21 measurements describing a truss network of four cells (Sec. 3.2) were taken on 99 specimens from the previous analysis: 47 from stream samples and 52 from the lake samples. A plot of the size-free sheared second principal component against component I (Fig. 4.6.11c) displays the same degree of discrimination between the two forms. In contrast to the convergent growth vectors of the atherinids (Sec. 4.6.1), the major axes of the clusters representing lake forms and stream forms diverge slightly with increasing size. When the loadings on the sheared second component are depicted directly on the truss network (Fig. 4.6.12), it is apparent that the primary difference in shape between the morphotypes is indeed one of robustness, with the stream form being deeper-bodied along the entire form. The contrast in loadings is between body-

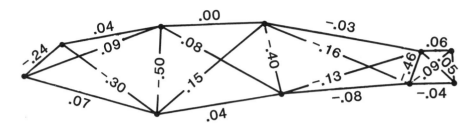

Figure 4.6.12 Loadings of the sheared second component of the truss data (Fig. 4.6.11c) depicted on the truss network. The greater in magnitude the loading, the more the forms differ along the corresponding truss element.

depth measurements, which have high negative values, and the more-or-less longitudinal measures, which have loadings near zero or slightly positive.

4.6.2.2 Shape differences between size-standardized forms.—We may visualize the same difference in shape between the stream and lake morphotypes by way of a biorthogonal analysis of the configurations of mid-sagittal landmarks, reconstructed from truss measures. The biorthogonal method compares pairs of single forms, usually individuals taken to be exemplars of populations. In the atherinid example of Section 4.6.1.2, for example, we selected "typical" individuals closest to the centroids of their clusters in character space. However, size-standardization of truss networks (Sec. 4.4) allows us to use composite averaged forms for biorthogonal analyses. Such averaged forms are preferable to the use of single representative individuals in describing differences in shape among populations in that most effects of individual variation have been averaged out explicitly.

The truss data for the two morphotypes of *C. cognatus* were standardized (within samples) to a common body size in two steps. First, we regressed the log-transformed measured distances on a composite measure of body size (the scores on the first among-group principal component; covariance matrix, log-transformed data). Second, we reconstructed the forms using the distance values predicted at a **standard body size** (in this case, corresponding approximately to a body length of 65 mm standard length). Because the size-standard chosen is approximately the average size for all individuals, we have produced averaged forms by reconstruction from the average measurements. For simplicity we join the configuration of landmarks by straight lines, forming polygons.

The biorthogonal analysis of these polygons (Fig. 4.6.13) shows again that the primary difference in shape is one of relative body-depth, with virtually no change in relative positions of fins and other structures and no differences in the change in shape between dorsal and ventral regions of the body. The dilatations along the longitudinal grid lines have values near 1.0, indicating no difference in relative body length at this standardized body size. Dilatations along the vertical grid lines show that the difference in robustness is primarily in the head, abdominal region, and peduncle. The

4.6 Allometric Growth and Shape-Change

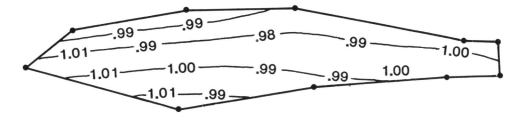

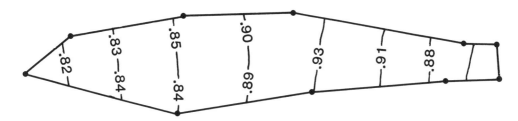

Figure 4.6.13 Biorthogonal analysis of averaged trusses for lake and stream sculpins. The dilatations describe the transformation from the stream morphotype to the lake morphotype.

forms are very similar in the region between the second-dorsal and anal fins. These results are concordant with the pattern of loadings on the sheared second principal component (Fig. 4.6.12).

4.6.2.3 Shape difference in relation to growth.—Divergence in form among conspecific populations might result from differences in the timing of specific developmental events, from altered patterns of local regulation during growth, or from some combination of these factors (see Sec. 5.2). To the extent that differences in postlarval allometry (estimated from samples of individuals) reflect differential development during the ontogenies of individuals, we should be able to account for ecophenotypic shape differences in terms of observed disparities in allometry with respect to general size.

The patterns of allometry of the lake and stream samples (Figs. 4.6.14a,b) differ markedly, both in the rate of development of body depth with respect to general size and in the relative growth rates of the medial-fin bases. The differences in allometric coefficients between the two forms (Fig. 4.6.14c) show the greatest between-group differences to be in relative rates of change of body-depth and of elongation of the caudal peduncle. These differences in loadings can be construed as loadings on another factor, one accounting for the effect of standard size upon the shape difference computed at that size. Considered as a deformation, this factor can be visualized directly by biorthogonal grids of its own, as demonstrated in Strauss and Bookstein (1982).

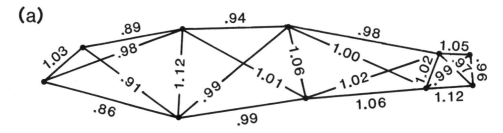

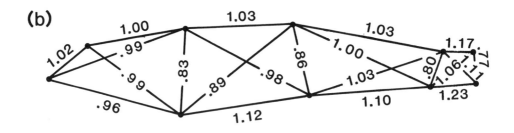

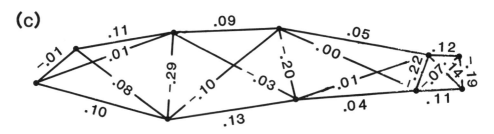

Figure 4.6.14 Multivariate within-group allometric coefficients, depicted on the truss networks. (a) Stream samples. (b) Lake samples. (c) Corresponding differences in coefficients between stream and lake morphotypes.

The allometric differences are generally concordant (Kendall's $\tau = 0.82$) with the pattern of loadings on sheared component II (Fig. 4.6.12). There are three notable disparities: the factor loadings do not show as large a contrast in peduncle length as is indicated by the relative allometries; the loadings describe a notable difference in head depth that is not paralleled by the allometric differences; and the differences in relative growth of the medial-fin bases do not result in a contrast between forms on the sheared second component. Nevertheless, a large proportion of the divergence in body form of lake and stream samples can be accounted for by observed differences in allometry.

4.6 Allometric Growth and Shape-Change

4.6.3 Shape differences in closely related species of *Cottus*

Throughout this book we have stressed the need for an adequate description and interpretation of shape-difference or shape-change with respect to biologically homologous landmarks on the form. The choice of landmarks is critical because only the biological homology of two configurations makes meaningful their scientific description or comparison. For a factor analysis, a rational choice of distance-measures among landmarks is important because the results depend upon the particular set of measurements chosen; if the measures taken do not accidentally or purposely represent the primary contrasts in form, the resulting analysis will be inadequate.

Although we originally formulated the truss method (Sec. 3.2) as a means of reconstructing a planar configuration of landmarks, we have found that the network systematically provides coverage of the form that increases the likelihood of uncovering shape differences among groups when the differences are not specific to a few local structures. We now describe a case in which use of a truss network in place of a conventional character set enhances discrimination between two morphologically similar species.

4.6.3.1 Discrimination between species.—*Cottus pitensis* and *Cottus klamathensis* are two species of western North American sculpins, sympatric throughout the range of *C. pitensis*, which are morphologically similar but highly variable in morphology both within and among populations. To test the ability of the truss network to discriminate these taxa, we measured 21 distances (describing a four-cell truss network) on 64 specimens, 21 of *pitensis* and 43 of *klamathensis*. For comparison we used a set of 19 conventional measures (Table 4.6.2) which had previously been taken on the same 64 individuals. Five of the 21 truss measures are shared with the conventional set.

Both sets of measurements were subjected to separate principal component analyses, each pooling the two species samples (Fig. 4.6.15). The truss data clearly provide better discrimination between the taxa. When the loadings on the sheared second component are depicted directly on the truss network (Fig. 4.6.16), they permit a geometrical interpretation in terms of the spatial arrangement of the distance measures. The primary contrasts on the form are between positive longitudinal loadings, which indicate that for a particular body size *pitensis* is relatively longer than *klamathensis*, and some of the negative diagonal and vertical loadings that show *klamathensis* to be relatively deeper-bodied than *pitensis*. In particular, the distances with highest negative loadings are posteroventrally oblique in the head and abdominal regions, and posterodorsally oblique on the peduncle. These principal directions of shape difference are largely unsampled by the conventional character set, except by the measures of maximum body depth and minimum peduncle depth, both of which have high negative loadings on the sheared component (Table 4.6.2). Thus the truss network improves shape discrimination by enforcing the use of diagonal measurements often not considered in conventional character sets.

Table 4.6.2 Principal component loadings of two different character sets for *Cottus klamathensis* and *C. pitensis*. Sample sizes are $n = 43$ and $n = 21$, respectively.

Conventional character set			Truss character set			
Character	I	Sheared II	Cell	Measure	I	Sheared II
			1	1-2	.21	.41
				1-3	.22	.21
				2-3	.23	−.20
				1-4	.22	.41
				2-4	.23	−.02
Maximum body depth	.24	−.18	2	3-4	.22	−.11
				3-5	.24	.24
				4-5	.23	−.13
				3-6	.23	.05
Length first dorsal fin base	.23	.14		4-6	.24	.20
			3	5-6	.21	−.10
Length anal fin base	.20	.14		5-7	.20	.26
				6-7	.21	.13
				5-8	.21	−.01
Length second dorsal fin base	.22	.19		4-8	.20	.18
			4	7-8	.23	−.29
				7-9	.19	.07
				8-9	.19	−.03
				7-10	.21	−.26
				8-10	.18	.03
Least depth caudal peduncle	.23	−.65		9-10	.23	−.42
Head length	.21	.05				
Postorbital head length	.22	.11				
Eye diameter	.17	−.19				
Premaxilla length	.25	.25				
Maxilla length	.26	.21				
Length first dorsal fin to caudal base	.24	.12				
Length second dorsal fin to caudal base	.23	.18				
Length anal fin to caudal base	.22	.11				
Length snout to pectoral base	.23	−.04				
Length pectoral fin margin to caudal base	.24	.08				
Length depressed second dorsal fin	.23	.28				
Length depressed anal fin	.22	.20				
Standard length	.21	.23				
Total length	.23	.28				

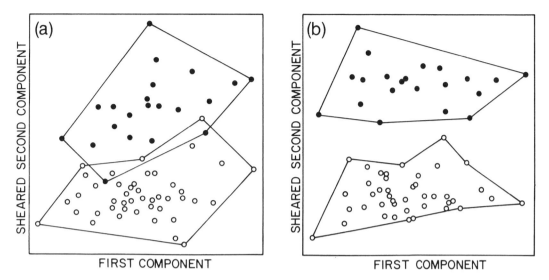

Figure 4.6.15 Principal component analyses of two character sets on the same 64 specimens of the sculpins *Cottus pitensis* and *C. klamathensis*. (a) Scatter plot of the sheared second component against the first among-group component of the traditional character set. (b) Scatter plot of the same components of the truss data set.

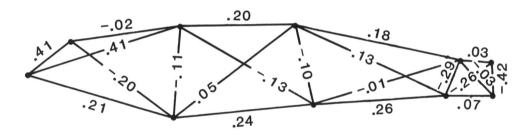

Figure 4.6.16 The differences in size-free shape between *Cottus klamathensis* and *C. pitensis*. Loadings of truss measures on the sheared component of Fig. 4.6.15.

4.7 The Components of Shape-Change

Current methods in numerical taxonomy and cladistics rely on measured lengths across forms or upon subjective evaluation of similarity and difference among shapes. Frequently, subjective classes of observed shapes or intervals of similar values of measured distances are represented in analysis by codes. Because of their subjectivity, evaluation and coding are often inadequate to represent shape information or analyze its differences. These purposes would be better served by more precise methods of shape analysis.

162   4 ANALYTIC METHODS

Because the analysis of shape *differences* is often the objective of comparative studies, it is appropriate to measure those differences directly, avoiding the intractable problem of shape measurement. For analysis of shape differences among a collection of fish neurocrania, we exploit the flexibility of the biorthogonal method to explore the complexity and the amount of shape change. By **complexity** we mean the heterogeneity of the amount and direction of change from region to region across the form. This heterogeneity is of interest as an indication of **regional dissociation** or independence of developmental trends among regions of the forms; such changes or differences are the products of genetic, ontogenetic, and evolutionary processes. Our goal thus becomes the quantification of shape differences in terms of the amount and complexity of the change required to transform one shape to another.

To make biological forms amenable to such studies, we first represent the shapes by sets of homologous landmarks (Sec. 1.2.1). From the deformations indicated by the landmark configurations of the different neurocrania we calculate the field of principal dilatations and directions, which describes, both locally and generally, the stretch and shrink of the shape change or shape difference (see, for instance, Fig. 4.7.9) for each pair of neurocrania. These local expansions and contractions can be large or small—extensive or subtle. Further, a pattern of change may be simple or complex independent of magnitude. Even a large change in shape, such as a large longitudinal extension, may be simple if it is relatively uniform over the entire area undergoing transformation. Alternatively, the dilatations may show considerable heterogeneity among regions of the forms.

We distinguish between two aspects of change: **scale-change**, at equal rates in all directions, and **proportion-change**, at different rates in different directions. These two aspects of change might have different genetic or epigenetic causes. The role of these causes in determining skeletal architecture, suture patterns, allometry, functional relations, etc., is still largely unstudied, not for lack of interest but for lack of suitable methods. The need for measures of dissociation among regions and of the relation between change in shape and change in size has been recognized for some time (Gould, 1977; see also Sec. 5.2).

4.7.1 Displacement and complexity

The information in a biorthogonal description of change is carried by three parameters at each mesh point: two dilatations and the single angle specifying the orientation of each cross upon one of the forms. Ignoring the angle, we can plot the two dilatations upon a Cartesian graph (Fig. 4.7.1), one coordinate of which is the logarithm (to any base) of the larger dilatation, the other coordinate the log of the smaller. Inverting the transformation—mapping the change in the reciprocal direction—results in dilatations that are the inverses of those for the original map (Bookstein, 1980a). In Figure 4.7.1, the inverse of the transform for the single cross is plotted at a different point, the reflection of the previous point in the line $\log d_1 +$

4.7 The Components of Shape-Change

$\log d_2 = 0$. Because the reflection preserves all distances, the geometry of the figure can eventually provide us with a fully symmetric measure relating two forms: an unsigned measure of shape difference.

Our two measures of form-difference—scale-change and proportion-change—are embodied in this plot. Consider the lines that bisect the coordinate axes in Figure 4.7.2. Their equations are $\log d_1 = -\log d_2$ (or $d_1 d_2 = 1$) and $\log d_1 = \log d_2$ (or $d_1/d_2 = 1$). Points on the line $d_1 d_2 = 1$ correspond to transformations that leave area invariant locally: they multiply distance in one direction by the same factor by which they reduce distance in the perpendicular direction. It is this type of transformation to which we refer above as proportion-change. For any point, the perpendicular distance to this diagonal is the logarithm of the factor $d_1 d_2$ by which the transformation alters local area. We call this component **size displacement**; it may be positive or negative.

Similarly, points on the line $d_1/d_2 = 1$, or $d_1 = d_2$, correspond to transformations that alter distance in all directions through a point by the same factor. The isotropic transformations leave shape unchanged, altering nothing but scale; they may be considered pure size changes, which we have referred to as scale-changes. No point in Fig. 4.7.2 can be below the line $\log d_1 = \log d_2$. The distance from any point to this diagonal is its **shape displacement**; it is always positive.

The point $(\log d_1, \log d_2)$ may be re-expressed using vector components along these two lines, the **size-change axis** and the **shape-change axis**:

$$(\log d_1, \log d_2) = ([\log d_1 + \log d_2]/2, [\log d_1 + \log d_2]/2) +$$
$$([\log d_1 - \log d_2]/2, -[\log d_1 - \log d_2]/2)$$
$$= (\log \sqrt{d_1 d_2}, \log \sqrt{d_1 d_2}) + (\log \sqrt{d_1/d_2}, \log \sqrt{d_2/d_1}). \quad (4.7.1)$$

Thus any transformation at a point is uniquely the product of the scale-change component and the proportion-change component. Should the transformation be inverted, the scale-change component is reflected, but the proportion-change component remains the same, because

$$\log d_1/d_2 = \log (1/d_2)/(1/d_1).$$

Note that the components of scale-change and proportion-change have been specified without any measurement of size or shape separately. The scale-change between a pair of forms equals the difference between their size measures if size is suitably defined.† The measure of proportion-change is not the difference between two proportion measures.

Because the diagonals are at 90°, they can be used as coordinate axes themselves (Fig. 4.7.3, a rotation of 45° of Fig. 4.7.2). We can re-express the preceding vector

---

† The size measure serving the purpose is 1/2 the log of area, as in Equation 4.7.1.

164   4 ANALYTIC METHODS

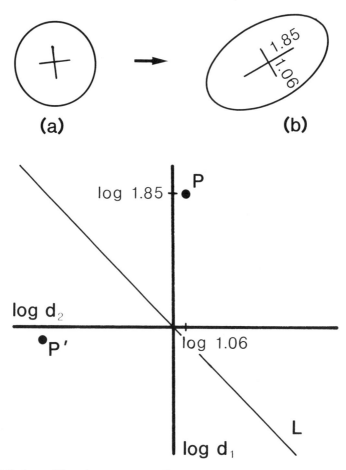

Figure 4.7.1 Log dilatations as coordinates. The plotted point represents the principal dilatations in the transformation of the circle (a) into the ellipse (b). The ordinate d_1 is the larger dilatation, the abscissa d_2 the smaller. Information about the orientation of the cross is not represented. P, point representing the transform from (a) to (b); P', point representing the transform from (b) to (a); L, line of reflection.

equation (Eqn. 4.7.1) as a scalar equation for the squared "length of the hypotenuse," a sum-of-squares decomposition:

$$(\log d_1)^2 + (\log d_2)^2 = \tfrac{1}{2} (\log d_1 d_2)^2 + \tfrac{1}{2} (\log d_1/d_2)^2 \qquad (4.7.2)$$

The left side of Equation 4.7.2 is the squared Euclidean distance of the point ($\log d_1$, $\log d_2$) from the origin ($\log 1$, $\log 1$) in the original coordinate system. The right side is proportional to that same squared distance in terms of the coordinates at 45° to the original axes (Fig. 4.7.3). By this identity we have decomposed a scalar quantity, a total change $(\log d_1)^2 + (\log d_2)^2$, into a pure scale-change component, $(\log d_1 d_2)^2/2$, and a pure proportion-change component $(\log d_1/d_2)^2/2$. This decomposition is the

4.7 The Components of Shape-Change

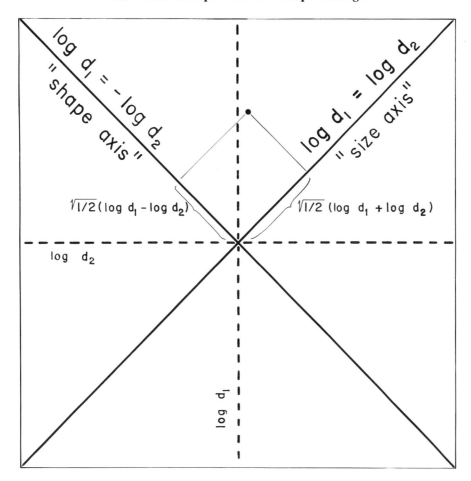

Figure 4.7.2 Scale- and proportion-change as Cartesian coordinates. One diagonal, the axis of scale-change ("size axis"), corresponds to transformations that leave all ratios invariant. The other diagonal, the axis of proportion-change ("shape axis"), corresponds to transformations that leave area invariant.

same for the inverse transformation, because reflection of a point in a coordinate axis leaves the squares of its coordinates unchanged.

In the rotated plot, the 45° lines correspond to a transformation having one dilatation equal to unity. Points on the 45° lines have a scale-change component and a shape-change component which are equal or opposite. If forms are standardized to constant length prior to transformation, then it is only an artifact that one of the dilatations is unity; therefore we use only the shape displacement for further analysis. A similar argument applies if standardization is by area. Whatever the form of standardization, dilatations for actual data vary around unity rather than being identical with it and are imperfectly confounded with shape displacement. In any

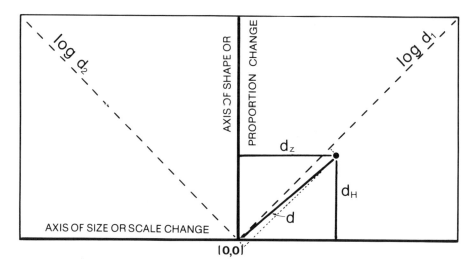

Figure 4.7.3 Squared Euclidean distances in the scale-proportion coordinate system. See text. Notice that the figure has been rotated by 45° from the previous figure.

case, the size displacement term should be ignored. When we standardize for size we discard the possibility of meaningfully analyzing size displacement.

4.7.1.1 Complexity as the scatter of a distribution.—If the dilatations were the same at every point of a transformation, that transformation would have the same shape-change and proportion-change at every point. In any extended biorthogonal grid, however, these components vary from point to point (Fig. 4.7.4). To each point on the interior of the form corresponds a point of the size-change—shape-change plane embodying the dilatations. Corresponding to the entirety of the transformation, there is covered an area of the size-change—shape-change plane. We might even imagine a distorted image of the form to indicate the general scheme of the mapping. We have a good sample of the points of the form: the evenly spaced collection of mesh points at which the biorthogonal algorithm computes its interpolation function. To this sample of points there corresponds a scatter in the size-change—shape-change plane, ordered (at least in the small) by a distortion of its grid on the form. Because d_1 is greater than d_2 by definition, all of the points fall above the scale-change axis.

Like any scatter, this one has a centroid and its own moments, or sums of squares, about that centroid. The coordinates of the centroid, in the system at 45°, give the **net size displacement** and **net shape displacement** of the transformation as a whole. Recall that the single point in Figure 4.7.3 bears a squared distance from (log 1, log 1) that is the sum of two components, scale-change squared and proportion-change squared (Eqn. 4.7.2). Likewise, the points of the scatter have two components of squared distance from their centroid, one for variance along the scale-change axis and one for variance along the proportion-change axis. The sum of these two variance components expresses the extent to which the observed change of form is heterogeneous, that is, the **complexity** C of the transformation.

4.7 The Components of Shape-Change

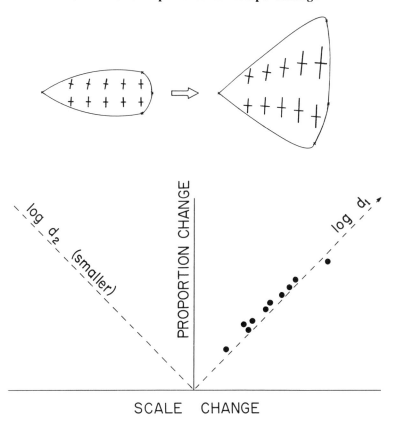

Figure 4.7.4 Scatter of dilatations for a nonlinear transformation. To each point on the interior of one form corresponds a point on the scale-proportion plane encoding the dilatations there. Note that the plotted points are not independent of one another.

The total sum of squares around (log 1, log 1) of the log dilatations of any transformation grid allows this decomposition into four parts: a mean scale-change component $\bar{Z}$ (size displacement), a mean proportion-change component $\bar{H}$ (shape displacement), a measure of scale complexity, and a measure of proportion complexity (Fig. 4.7.5). Of the four components, the last three are invariant against change in scale in the geometric forms separately.† The two complexity measures can be further decomposed into sums of squares within body regions and between body regions.

---

† The analogy between these components and ordinary statistical moments breaks down around the line $d_1 = d_2$, past which the data may not trespass. Whenever the biorthogonal grid includes a singularity, the scatter in the (log d_1, log d_2)-plane will include a sector that has been folded back over this line, so to speak. Because singularities are isolated, the continuous image of which our scatter is the sample will touch the line at only one point.

168   4 ANALYTIC METHODS

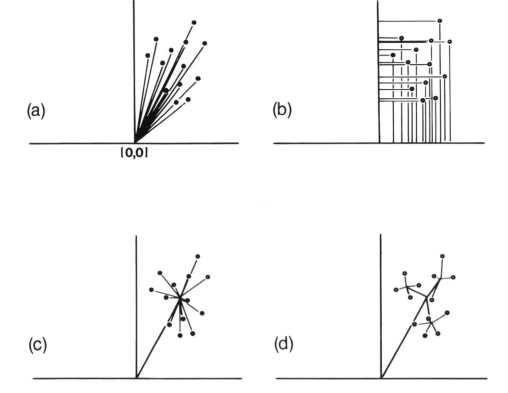

Figure 4.7.5 Decomposition of sums of squares in the scale-proportion plane. (a) Displacement of individual points from the origin (log 1, log 1). (b) Scale-and proportion-displacement of individual points. (c) Mean displacement and the scatter about the mean. The variance of this scatter is a measure of the complexity of the transformation. (d) Scatters about mean displacement computed separately within several regions.

4.7.2 Catostomid skulls

In the following examples, we use indices of shape difference among neurocrania of several genera of catostomid fishes to evaluate displacement and complexity in comparisons that illustrate phyletic divergence and parallelism among genera. The fishes of the family Catostomidae are classified in 13 genera whose phylogeny has been extensively studied. Shape differences among neurocrania have not been explicitly used in their classification, except when expressed as the presence or absence of the supraorbital bones and the several fontanelles. A cladogram of catostomid genera is shown in Figure 4.7.6. Exemplars of these genera were selected for biorthogonal analysis by choosing representative specimens of several sizes from available neurocrania. Photographs of dorsal views of the (relatively flat) neurocrania were enlarged to constant length (from the base of the median process of the dermethmoid

4.7 The Components of Shape-Change

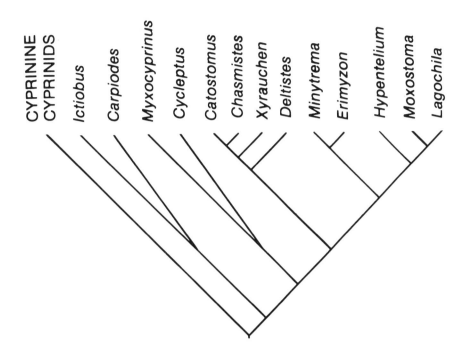

Figure 4.7.6 Cladogram of catostomid genera. (Based on data of G. R. Smith, unpublished.)

to the base of the median process of the supraoccipital) and homologous landmarks were identified using the criteria in Chapter 1. The landmarks were digitized to provide outlines for biorthogonal analysis (Figs. 4.7.7,8).

4.7.3 Reading a complexity plot

A large shape difference between neurocrania is illustrated in Figures 4.7.9a,b. The crosses in (a) depict the directions and dilatations quantifying the distortion by which a 3 cm skull of *Hypentelium* differs, after standardization of length, from a 2.7 cm skull of *Erimyzon*. The anterior one-third of *Hypentelium* is disproportionately expanded in several directions; the posterior part of the skull is shorter than in *Erimyzon*. These large differences correspond to those points in (b) that are far from the origin. Few of the points are near the horizontal axis; this means that there are few

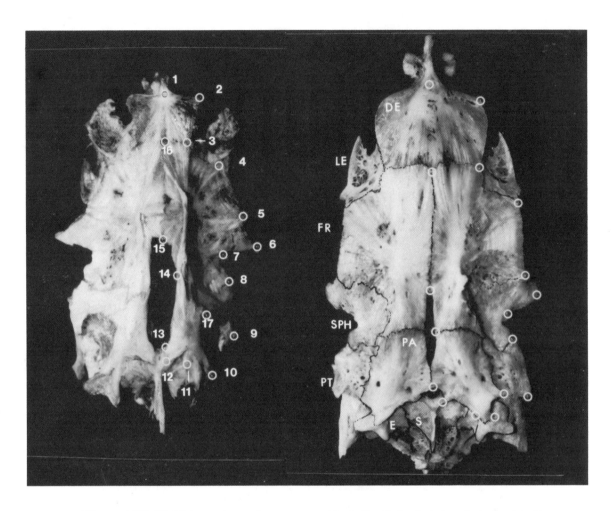

Figure 4.7.7 Digitizing scheme for catostomid skulls. Left, *Ictiobus bubalus* (skull length 19 mm); right, *Deltistes luxatus* (87 mm). Landmarks: (1) base of dermethmoid anterior process, (2) anterior corner of dermethmoid, (3) posterolateral corner of dermethmoid, (4) origin of frontal:lateral-ethmoid suture, (5) terminus of frontal:sphenotic suture, (6) posterolateral corner of sphenotic wing, (7) notch at base of sphenotic wing, (8) terminus of sphenotic:pterotic suture, (9) posterolateral corner of dermal roof of pterotic, (10) posterolateral corner of parietal, (11) junction of sutures among parietal, epiotic, and supraoccipital, (12) posteromesial corner of parietal, (13) terminus of parietal:supraoccipital suture at fontanelle, (14) terminus of parietal:frontal suture at fontanelle, (15) terminus of frontal suture at fontanelle, (16) terminus of frontal suture at dermethmoid, (17) interior point—terminus of pterotic parietal suture in posttemporal fossa. Abbreviations: *DE* dermethmoid, *LE* lateral ethmoid, *FR* frontal, *SPH* sphenotic, *PA* parietal, *PT* pterotic, *E* epiotic, *S* supraoccipital.

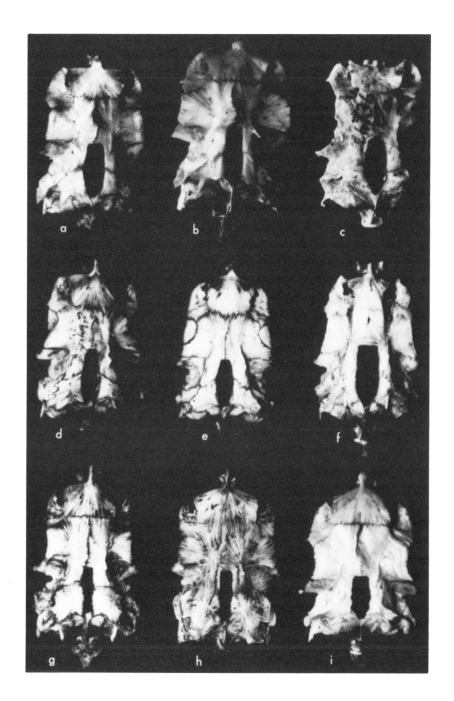

Figure 4.7.8 Catostomid skulls. (a) *Moxostoma valenciennesi* (skull length 22 mm); (b) *Moxostoma valenciennesi* (60 mm); (c) *Lagochila lacera* (16 mm); (d) *Minytrema melanops* (26 mm); (e) *Minytrema melanops* (57 mm); (f) *Erimyzon sucetta* (27 mm); (g) *Catostomus tahoensis* (45 mm); (h) *Chasmistes cujus* (76 mm); (i) *Hypentelium nigricans* (34 mm).

areas of little distortion over the transformation (compare with Fig. 4.7.10). There is wide scatter along the scale component, with somewhat separate clusters of points in the upper left and upper right parts of (b), representing large local decreases (left) and increases (right) in scale.

Using this information we may examine the variation in shape change over different regions of the neurocrania. Recall that variability in the dilatations, among and within regions, provides measures of the complexity of the transformation. Notice in particular the regional variability of the shape differences among the neurocrania—low variation in the dilatations within regions and large variation among regions. Because the bones of interest can be viewed as a longitudinal series (Fig. 4.7.7), the plots of scale- and proportion-differences against longitudinal position on the skull show profiles of complexity among and within regions (Figs. 4.7.9c,d). Each of these plots shows displacement and complexity, column by column. Displacement is the distance from the zero line to the central tendency of a column or a group of columns; the complexity of each column or group is its vertical scatter. **Scale-complexity** and **proportion-complexity** are the variances around mean displacement over all columns on the plot; regional complexity is the vertical scatter within regions.

The longitudinal plot of scale-difference (Fig. 4.7.9c) shows high positive values for the five columns representing the anterior (dermethmoid) part of the neurocrania. The next eight columns of values are variable, with column medians descending posteriorly. The last eight columns represent the sphenotic and occipital regions by points that are generally low. We interpret (c) to show a large isotropic expansion (expansion by nearly the same factor in every direction) of the dermethmoid of *Hypentelium* relative to *Erimyzon*. The posterior parts of the frontal region and especially the sphenotic and occipital regions of the *Hypentelium* skull are relatively reduced in scale. Proportion-differences between *Hypentelium* and *Erimyzon* are as variable within regions as the scale-differences, but less variable among regions (Figs. 4.7.9c,d). The outlying points on these plots, such as the point in column 7 of Figure 4.7.9c that shows highly negative scale-difference in the frontal region of the skull, are artifacts caused by extreme kinks along the boundary.

Variability along the profiles of scale-difference and proportion-difference reveals the extent to which the difference in skull shape is heterogeneous, or regionalized. As illustrated in Figures 4.7.9b–d, the difference between the exemplars of *Erimyzon* and *Hypentelium* is complex, showing dissociation among regions. The anterior part of the *Hypentelium* skull is larger and wider; the difference is not an extension of a pattern seen elsewhere in the skulls. The large, complex differences in skull shape suggest that *Hypentelium* and *Erimyzon* have diverged considerably.

A large shape change that is simple is illustrated in Figure 4.7.10a–d. Skulls of the catostomin genera *Chasmistes* and *Deltistes* differ by a relatively simple, or homogeneous, transformation, greatly altering the proportion of length to width. The amount of divergence is less than that in the previous example, despite the obvious displacement in shape (Fig. 4.7.10c,d). The scatter of proportion difference against longitudinal position (Fig. 4.7.10d) illustrates dissociation among regions of the skull.

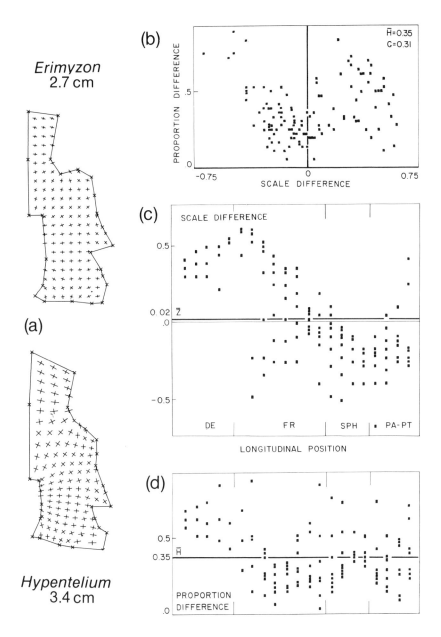

Figure 4.7.9 Shape comparison of *Erimyzon* and *Hypentelium* skulls. (a) Outline of right side of skull roofs of *Erimyzon* and *Hypentelium*. Biorthogonal crosses show the principal directions of the transformation. (b) Scatter of logarithms of the dilatations (ratios of homologous cross arms) on the shape (proportion) and the size (scale) components. (c) Scatter of scale displacement against longitudinal position on the skull. Differences in vertical scatter are indications of complexity. (d) Scatter of proportion displacement against longitudinal position on the skull. Abbreviations: $\bar{H}$, mean shape displacement; C, complexity—the variance of the scatter around its centroid; $\bar{Z}$, mean scale displacement; de, dermethmoid region of the skull; fr, frontal region; sph, sphenotic region; pa-pt, parietal-pterotic region. Skull length is given in centimeters.

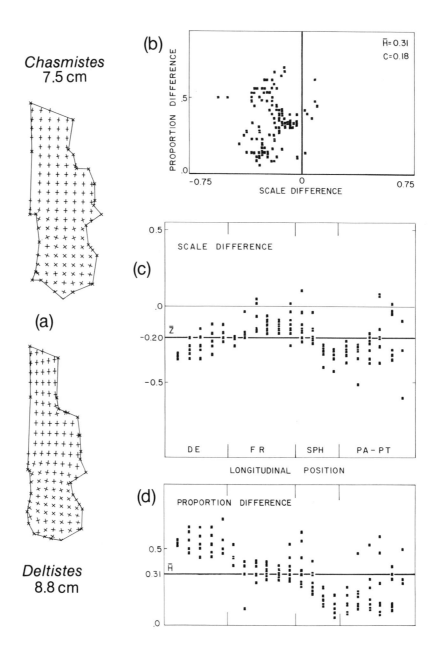

Figure 4.7.10 Shape comparison of *Chasmistes* and *Deltistes* skulls. (a) Outline of right side of skull roofs of *Chasmistes* and *Deltistes*. Biorthogonal crosses show the principal directions of the dilatations. (b) Scatter on shape (proportion) and size (scale) components. Low complexity is indicated by the tight scatter. (c) Scatter of scale displacement against longitudinal position on the skull. (d) Scatter of proportion displacement against longitudinal position on the skull. The scatter shows dissociation among regions of the skulls. Abbreviations are as in Figure 4.7.9.

4.7 The Components of Shape-Change

We have introduced the size and shape displacements of the transformation as two different quantities. One possible net measure of the difference between two forms would be the distance—termed the mean displacement—of the centroid of the points in (b) from (log 1, log 1) (see Fig. 4.7.5c). For the transformation between *Deltistes* and *Chasmistes* the mean displacement is 0.33, as compared to 0.35 between *Erimyzon* and *Hypentelium*. But because we have standardized by length, size displacement is not meaningful, and therefore mean displacement is also misleading. Consequently, it is more appropriate to compare mean shape displacements: 0.30 for *Deltistes* and *Chasmistes*; 0.35 for *Erimyzon* and *Hypentelium*. The complexity (the mean distance of points from their centroid in Fig. 4.7.10b; see also Fig. 4.7.5c) between *Deltistes* and *Chasmistes* is 0.14; the complexity between *Erimyzon* and *Hypentelium* is 0.31. Thus, the latter shape change can be characterized as large and complex; the former is characterized as large but simple.

An example of a modest change in shape is seen in the difference between a 22 mm skull of *Moxostoma valenciennesi* and a 60 mm skull of the same species (Fig. 4.7.11a–d). Most of the log dilatations are near (log 1, log 1). The shape displacement is 0.18 (compared to 0.35 between *Erimyzon* and *Hypentelium*) and the complexity is 0.16 (compared to 0.31 between *Erimyzon* and *Hypentelium*). The scale- and proportion-changes are relatively homogeneous (Fig. 4.7.11c,d) except in the corners and at the posterior edge.

4.7.4 Phylogenetic applications

We now extend the methods developed above to a study of similarity of skull shapes in relation to cladistic hypotheses. For example, let us compare *Moxostoma* (redhorse suckers), *Hypentelium* (hogsuckers), *Lagochila* (harelip sucker), *Erimyzon* (chubsuckers), *Minytrema* (spotted sucker), *Catostomus* (finescale suckers) and an outgroup, *Ictiobus* (buffalo suckers) (Figs. 4.7.6–8).

Seven neurocrania of adults (and seven neurocrania of smaller specimens listed here preparatory to later analysis) are compared as described earlier: *Ictiobus bubalus* (skull length, from base of dermethmoid process to base of supraoccipital process, 19 mm), *Minytrema melanops* (26, 57 mm), *Catostomus tahoensis* (19, 45 mm), *Moxostoma valenciennesi* (22, 60 mm), *Erimyzon sucetta* (12, 27 mm), *Hypentelium nigricans* (16, 34 mm), *Lagochila lacera* (16 mm), and *Moxostoma duquesnei* (22, 50 mm). Interspecific comparisons involving *Moxostoma* always use *Moxostoma valenciennesi*.

Among seven adults in the first analysis there are 21 transformations (and their inverses) representing shape differences accumulated over evolutionary time. Branch lengths of a cladistic hypothesis may imply relative amounts of divergence among taxa. Therefore, the amounts of shape differences observed between pairs of taxa and some outgroup should be correlated (however imperfectly) with the relative amounts of time over which shared and unshared differences have accumulated.

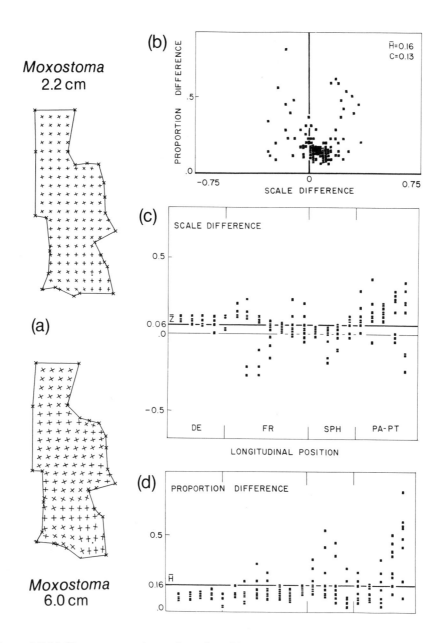

Figure 4.7.11 Shape comparison of small and large *Moxostoma* skulls. (a) Right side of skull roofs, showing little change. (b) Scatter on the shape (proportion) and the size (scale) components. (c) Scatter of scale displacement against longitudinal position on the skull. (d) Scatter of proportion displacement against longitudinal position on the skull. The scatters of (c) and (d) illustrate dissociation among skull regions. Abbreviations are as in Figure 4.7.9.

4.7 The Components of Shape-Change

We use shape comparisons for evolutionary inference, as did Lull and Gray (1949), who compared skulls of ceratopsian dinosaurs by distortion of Cartesian grids. They showed the phylogenetic utility of two kinds of comparisons: (1) similarity of transformations of two forms from an ancestral form, which implies recency of common ancestry, and (2) similarity of ontogenetic transformations, which implies genetic similarity and relationship.

As an example of the elements that enter into a comparison of type (1), consider (region by region) the lengths, ratios, and directions of the cross arms as well as the homogeneity of those parameters over the forms in the transformation from *Ictiobus* to the other taxa in Figure 4.7.12. The *Ictiobus-Catostomus* and *Ictiobus-Moxostoma* comparisons appear most similar. As an example of comparison type (2) (Fig. 4.7.13), the ontogenetic transformation of the two species of *Moxostoma* are most similar. The ontogenetic transformation in *Erimyzon* is similar to that between ontogenetic stages of *Minytrema*, but less similar to those between ontogenetic stages of *Moxostoma* or *Catostomus*. The pattern of the transformations implies more shared evolutionary history and thus less divergence between *Erimyzon* and *Minytrema* than between *Erimyzon* and the other forms.

More objectively, Table 4.7.1 presents the displacements and complexities among six taxa whose relative similarity to the outgroup is of interest. *Hypentelium* is most displaced from the outgroup and differs by the most complex transformation ($\bar{H} = .68$, $C = .42$), thus polarizing the transformation series among these shapes. *Catostomus* differs least from *Ictiobus* ($\bar{H} = .42$, $C = .34$); *Minytrema* differs by a similar amount ($\bar{H} = .48$, $C = .33$) and is also most similar to *Catostomus* ($\bar{H} = .22$, $C = .17$). *Moxostoma* ($\bar{H} = .28$, $C = .15$) and *Hypentelium* ($\bar{H} = .26$, $C = .21$) are likewise similar to *Catostomus*.

Lagochila, a sister group of part of *Moxostoma*, is unexpectedly similar to *Catostomus* ($\bar{H} = .33$, $C = .39$), but its transformations to several more closely related species are less complex. Although it differs substantially from adult *Moxostoma* ($\bar{H} = .44$, $C = .36$), it is more similar to a young *Moxostoma* (skull length 2 cm; $\bar{H} = .35$, $C = .30$), suggesting that its skull shape has been affected by the process of paedomorphosis.

Let us consider the right-hand branch of the cladogram (Fig. 4.7.6), in which *Erimyzon* is postulated to have had a most recent common ancestry with *Minytrema*. (However, *Erimyzon* shares synapomorphies with genera in the *Moxostoma* and *Catostomus* groups as well.)

The form-differences between *Minytrema* and *Erimyzon* are intermediate. The shape displacement of the transformation is 0.31; its complexity is 0.27 (Table 4.7.1). The primary axis of compression is anteromesial to posterolateral, through the dermethmoid and frontals. The transformation from *Moxostoma* to *Erimyzon* also involves decreased width across the front of the skull. The amounts of shape difference (0.39) and shape complexity (0.24) are similar to those of the previous example. The transformation between *Catostomus* and *Erimyzon* has a shape difference index of 0.35 and a shape complexity of 0.26. Because the neurocranium of *Erimyzon* is only slightly more similar to *Minytrema*, the cladogram remains relatively unchanged.

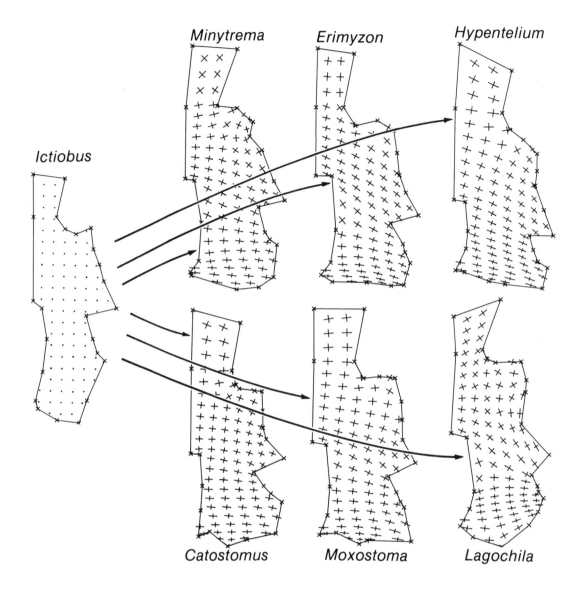

Figure 4.7.12 Comparison of forms of sucker skulls with an outgroup form. Biorthogonal comparisons from *Ictiobus bubalus* to six relatives of *Moxostoma* and *Catostomus*. All of the derived forms show posterolateral broadening across the frontal bones as well as longitudinal shortening of the back of the skull. *Minytrema*, *Erimyzon*, and *Hypentelium* bear similar principal directions and dilatations except anteriorly. *Moxostoma* and *Catostomus* are similar because the two transformations are nearly isometric. There are no crosses on the form of *Ictiobus* because they lie differently for each of the six comparisons.

4.7 The Components of Shape-Change

The values in Table 4.7.1 suggest ways in which shape differences can be quantified for character analysis. Displacement and complexity are used to classify the shapes (with respect to the outgroup) into character states. As a first step, if complexity is high, subdivision of the anatomical unit might be warranted, or a more general index can be calculated. For the present example, we divided the neurocranium into four regions—dermethmoid, frontal, otic, and occipital—on the basis of apparent discontinuities in the trends of displacement and complexity along the length of the skull. Once this is done, a convenient summary of the 12 dissimilarity measures (shape and size complexity and shape displacement in each of the four regions) is the first principal component of their measure-by-measure correlation matrix. This summary index is sensitive to the character's (i.e., the skull's) divergence or convergence among the taxa, but is less sensitive to local details.

Although these measures quantify differences among homologous structures, individually they need not reliably indicate recency of common ancestry. Let the branching sequence of four taxa be represented by a cladogram with inter-taxon "distances" b_{12}, b_{13}, b_{14}, b_{23}, b_{24}, b_{34}. These four taxa can also be represented in character space as points on a tetrahedron with six edges of lengths e_{ij}. Because the tetrahedron may take a variety of shapes, we can expect the lengths b_{ij} to be positively but imperfectly correlated with the lengths e_{ij}. (See Sec. 5.4.1.) The correlation will be perfect only when evolution of shape differences has been constant in rate and uniformly divergent (Farris, 1979)—cladistic trees constructed from "distances" are unreliable unless these criteria have been met. Furthermore, differences, whether measured by immunology, DNA hybridization, or biorthogonal displacement and complexity, yield "distances" that do not ordinarily form a metric—one edge of a triangle (on the tetrahedron) may exceed the sum of the lengths of the other two. Yet the lengths e_{ij} are of interest because they describe the net amount of evolution in structures. Ordination or clustering of the skull data, as suggested in the previous paragraph, produces a transformation series or a shape-state tree whose values or branching sequence could be used in a cladistic analysis.

But evolutionary history cannot be reconstructed solely from differences among adults. Observed differences among descendants of a common ancestor result from changes in ontogeny that must be considered in reconstruction of relationships. In the above examples we considered shape differences as static problems in cladistics. We now add young stages to the analysis in order to develop a simplified example of a dynamic approach.

4.7.5 Shape comparison of ontogenies

To illustrate examples of ontogeny in relation to phylogeny, after the spirit of Lull and Gray (1949), we compare a more extensive sample of catostomid neurocrania using a slightly different set of landmarks, internal as well as external. In order to describe the relationships between shape differences, independent of size, we analyze

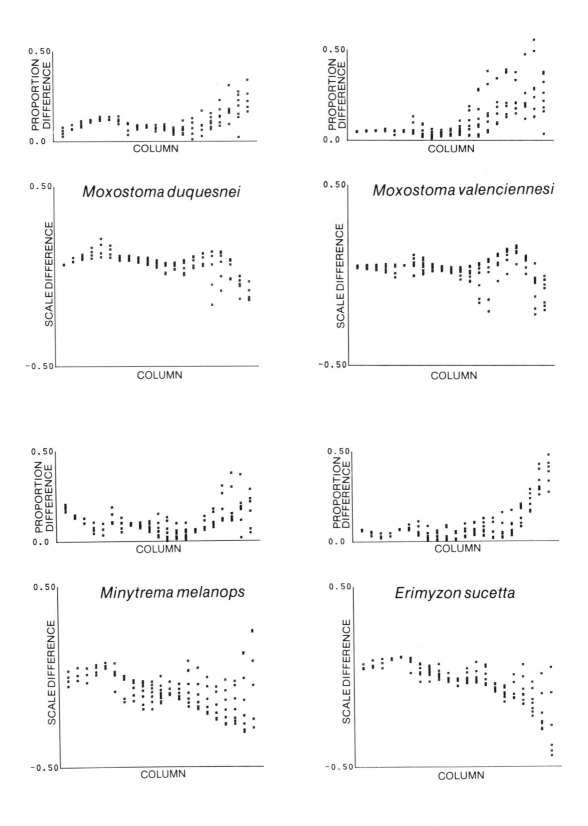

4.7 The Components of Shape-Change 181

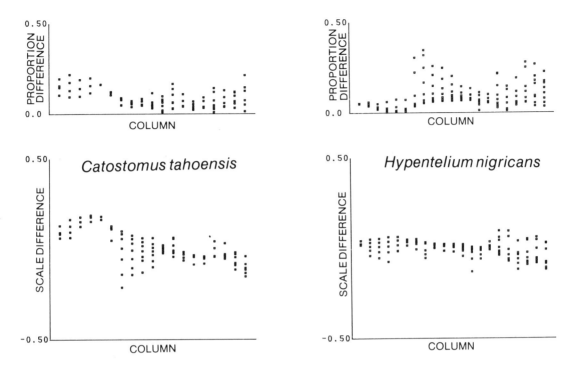

Figure 4.7.13 Comparison of ontogentic transformation among *Erimyzon* and possible sister groups. Similarity of any of these transformations to the shape change in *Erimyzon* might provide evidence of relationship.

neurocrania of different sizes using only shape displacement and complexity. Although the measures are not metric, they reveal divergence or convergence of growth trajectories when taxa are compared two at a time.

Two ontogenetic series, each sampled at two stages, can be represented by six values for shape dissimilarity among the four points: (1, 2) small to large within species, (3, 4) small to small and large to large between species, and (5, 6) small to large between species (Table 4.7.2). The relative similarity of pairs of growth trajectories are described by the between-lineage dissimilarity values.

Comparison of mixed size pairs among *Erimyzon*, *Hypentelium*, *Minytrema*, *Moxostoma*, and *Catostomus* are shown in Figures 4.7.14 and 4.7.15. Among these comparisons, two species of *Moxostoma* are most similar, as expected. Searching the matrix for similar comparisons involving each genus, we see that *Moxostoma*, *Minytrema*, and *Hypentelium* are each most similar to *Catostomus*, and *Erimyzon* is most similar to *Minytrema* and *Moxostoma*. These similarities in neurocranium shape are discordant with the shape differences on which the cladogram is based. The general similarity in skull proportions among medium-sized stream fishes—*Catostomus*, *Moxostoma*, *Minytrema*—may indicate that stream ecology has played a role in constraining shape changes. The cross-lineage comparisons of juveniles and adults

Table 4.7.1 Comparisons of shape differences among skulls of catostomid fishes. $\overline{H}$, proportion displacement; C, mean complexity of differences in skull shape, calculated according to methods described in the text.

	Ictiobus	Catostomus	Minytrema	Erimyzon	Hypentelium	Moxostoma
Catostomus	$\overline{H}=.42$ $C=.34$					
Minytrema	$\overline{H}=.48$ $C=.33$	$\overline{H}=.22$ $C=.17$				
Erimyzon	$\overline{H}=.64$ $C=.35$	$\overline{H}=.35$ $C=.26$	$\overline{H}=.31$ $C=.27$			
Hypentelium	$\overline{H}=.68$ $C=.42$	$\overline{H}=.26$ $C=.21$	$\overline{H}=.34$ $C=.23$	$\overline{H}=.35$ $C=.31$		
Moxostoma	$\overline{H}=.57$ $C=.33$	$\overline{H}=.28$ $C=.15$	$\overline{H}=.38$ $C=.21$	$\overline{H}=.39$ $C=.24$	$\overline{H}=.39$ $C=.24$	
Lagochila	$\overline{H}=.58$ $C=.37$	$\overline{H}=.33$ $C=.39$	$\overline{H}=.41$ $C=.41$	$\overline{H}=.50$ $C=.26$	$\overline{H}=.45$ $C=.44$	$\overline{H}=.44$ $C=.36$

reveal asymmetries in the patterns; for example, *Minytrema* becomes more similar to *Catostomus* as they grow, as expected in convergence. By contrast, the similarity of *Hypentelium* to *Moxostoma* and possibly *Erimyzon* to *Minytrema*, may be due to relative recency of shared ancestry.

Comparison of Figures 4.7.14,15 with the cladogram in Figure 4.7.6 invites interpretation of ontogenetic changes with respect to their outgroups. A large or complex difference between juveniles and adults of one species may indicate ontogenetic change comparable to terminal addition. (Variation in juvenile size affects the apparent change. In the example, Figures 4.7.14,15, *Catostomus* shows a small but complex change.) A small difference between juveniles and adults could mean either that growth in that interval was isometric or that differences have been deleted or truncated. In this example, *Erimyzon* adults differ little from juveniles because maturity is attained at a small size and the fish remain dwarfed.

The relative shape similarities of young and adult stages of two taxa suggest two sorts of questions: about the timing of shape change relative to growth stage and about the cladistic differences between taxa. Either can be answered better if the other is known (Fink, 1982). The logical manipulation of these known or inferred or guessed propositions requires some methodological principles. We choose the principle of parsimony so that explanations invoke the minimum number of changes necessary to account for the similarities and differences.

4.7 The Components of Shape-Change

Table 4.7.2 Ontogenetic comparisons of sucker skulls. Roman entries, proportion displacement; italicized entries, mean complexity, computed according to methods explained in the text. Each taxon is represented by two individuals. Skulls lengths (in cm) are indicated in the headings.

Taxon, skull lengths	*Minytrema* 2.6	*Minytrema* 5.7	*Catostomus* 1.9	*Catostomus* 4.5	*Moxostoma* 2.2	*Moxostoma* 6.0	*Erimyzon* 1.2	*Erimyzon* 2.7	*Hypentelium* 1.6	*Hypentelium* 3.4
Minytrema 5.7			15 / *16*	14 / *11*	20 / *29*	17 / *20*	18 / *41*	17 / *19*	15 / *19*	19 / *22*
Minytrema 2.6		10 / *16*	20 / *33*	19 / *30*	21 / *52*	16 / *19*	18 / *28*	19 / *24*	19 / *37*	21 / *53*
Catostomus 4.5					10 / *14*	14 / *19*	19 / *46*	20 / *26*	10 / *13*	14 / *15*
Catostomus 1.9				8 / *19*	14 / *14*	18 / *19*	19 / *35*	20 / *21*	15 / *22*	18 / *26*
Moxostoma 6.0							10 / *27*	20 / *17*	16 / *14*	16 / *26*
Moxostoma 2.2						12 / *9*	18 / *41*	16 / *19*	14 / *14*	13 / *14*
Erimyzon 2.7									19 / *28*	17 / *33*
Erimyzon 1.2								7 / *11*	18 / *41*	17 / *54*
Hypentelium 3.4										
Hypentelium 1.6										10 / *9*

Inset sub-table:

		Mox. v. 2.2	*Mox. v.* 6.0	*Mox. d.* 2.2	*Mox. d.* 6.0
Mox. v.	6.0				
	2.2		12 / *9*		
Mox. d.	6.0	12 / *25*	7 / *3*		
	2.2	11 / *8*	14 / *7*	10 / *7*	

Some of the possible combinations of data on relative similarities of young and adult growth stages of two taxa are as follows.

1. If the young of a sister pair were similar to each other and the outgroup, the adults different, and only one adult substantially different from its young and the outgroup, we could conclude that only the different adult has diverged, by ontogenetic changes that are manifested late in development. (We have no examples of this.) When both adults differ from their young and the outgroup, as well as each other, we conclude that both have diverged (e.g., *Moxostoma* and *Hypentelium*).

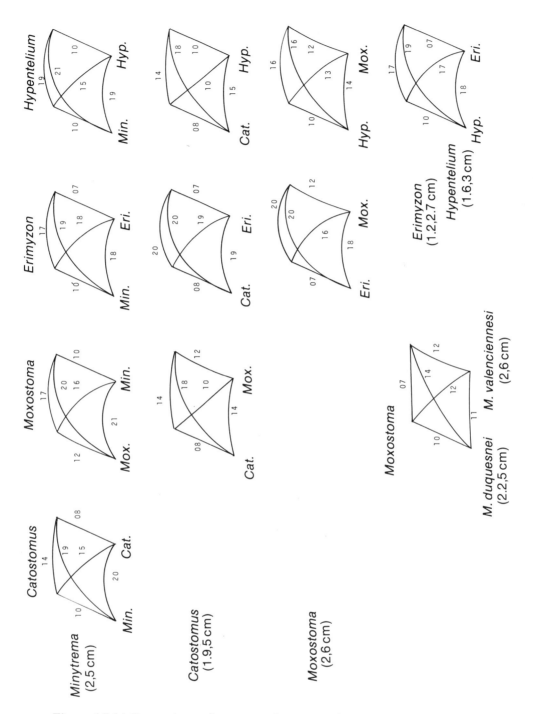

Figure 4.7.14 Comparison of amounts of ontogenetic and phyletic change. The tent diagrams graphically show the amounts of shape change between small (bottom corners) and large (top corners) skulls of suckers. Because the measure is not metric, the amounts of proportion-displacement (from Table 4.7.2) are represented by curved lines in space, the four points compared being arbitrarily fixed. The six values for each set of comparisons between two taxa, when observed in the context of Figure 4.7.6, give a sense of the relationship between ontogeny and phylogeny of the taxa.

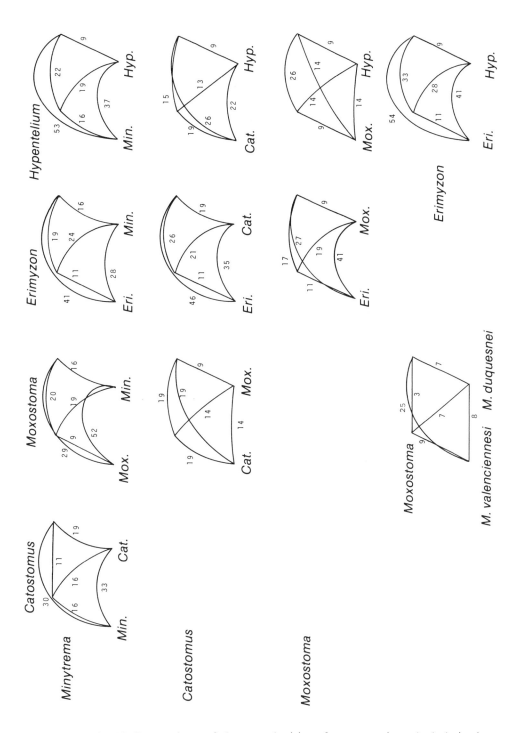

Figure 4.7.15 Comparison of the complexities of ontogenetic and phyletic change. The complexities of shape change between small and large skulls of pairs of sucker taxa are represented by values from Table 4.7.2, arranged in tent diagrams as explained in the previous figure.

2. If the pattern of similarities were as in (1) but the taxa were not sister groups, one might consider the possibility of retained plesiomorphy or convergent similarity of young stages, depending on their similarities to the outgroup. We have seen no examples of the latter—it might be rare and in any case it would be ontogenetically homoplastic.

3. Large change in a growth stage earlier than the two being compared usually results in differences that are preserved at all later stages (e.g., displacement between *Catostomus* and *Erimyzon*). This case is frequently seen in distant relatives. Its demonstration between relatives closer than those we have studied might be taken as an example of "macroevolution."

4. If adults are similar to each other and young are similar to each other, then parallel changes are occurring throughout development (e.g., *Moxostoma* and *Catostomus*). Frequently this results in similarity between the adults of one species and the young of the other. Large *Catostomus* are similar to young *Moxostoma* and *Hypentelium*, suggesting that the last two genera or their common ancestor share a derived ontogenetic shape change.

5. When older stages are more similar than young stages in both displacement and complexity, as in the comparison of *Minytrema* and *Catostomus*, convergence might be inferred—assuming that they are not sister groups that have differentiated only in younger stages.

In summary, the role of cladistic evidence in the above inferences can be useful, though its application becomes more ambiguous the more distant the relationships. If the juvenile and adult shapes are similar in sister groups, it is parsimonious to presume that divergence has not occurred. Among distant relatives, derived similarities in adults coupled with derived differences in young imply early ontogenetic and phylogenetic differentiation and convergence of late ontogenetic stages (e.g., *Catostomus* and *Minytrema*). If the adult shapes are primitive, convergence later in ontogeny is implied; if one of the juveniles bears the primitive form, convergence is still involved, but the interpretation varies according to what the adults are similar to. In all of these cases we would expect complexity to be relatively high.

In this section we have developed methods for quantifying shape change in the study of phylogeny and ontogeny. The methods may be used to quantify character-state trees or transformation series based on shape differences. Shape change is seen to have two separable elements, amount and complexity. In Section 5.2 we will speculate about extending these concepts to the analysis of developmental dissociation and evolution.

5

Geometric Morphometry and Evolutionary Biology

Our survey of the morphometric methods we prefer is now completed. We have explained how we record data and summarize them into descriptors of form and form-change; we have shown models for covariation of those descriptors and have demonstrated how to discern patterns in the data that fit those models. In our view, these expositions not only exemplify modern technique but also speak clearly to the larger role of morphological data in evolutionary biology. Beyond their ability to answer conventional biometric questions, we intend these methods to alter the semantics of the questions evolutionary biologists ask.

In this final chapter, which is somewhat speculative, we apply our methods to the terms and tactics of empirical evolutionary research. Much more than conventional biometry, geometric morphometry shares with evolutionary biology the grand subject of *change* or *difference* as the object of discourse. Realizing D'Arcy Thompson's suggestion that shape change be a mensurand in its own right, we have enabled the investigator to explicitly compute morphological comparisons as geometric transformations. Conventional biometric analyses of variance or covariance among characters selected a priori may be replaced by the study of ontogenetic or evolutionary histories depicted directly over the forms. To the extent that the stuff of evolutionary science is form-comparison, we have supplied additional methods for studying its questions directly, without the necessity of a character set intervening.

Researchers who would exhaustively exploit the geometry of form should eschew conventional notions of the morphometric "character set," the ad hoc motley of variables. The scores by which comparisons will ultimately be reported—the poles of a discriminatory factor, or the ratio of two arms of a principal cross—need not have been measured at all. They may instead have been *deduced* from the analysis of data gathered under the guide of an empirical point-homology function. The contrasts by which evolutionary hypotheses are tested generate variables by which the forms undergoing comparison may be optimally compared. Those variables are different for different contrasts, different questions. Within a single data base, the descriptors of ontogeny, for instance, may be quite different from the descriptors of phylogeny (Bookstein, 1978a:144–153). There results a new sort of parallelism of vocabulary (Sec. 5.3); evolutionary questions which used to be interpreted by way of characters may now be asked of the forms directly, using the language of form-change.

In this chapter we argue the import for empirical evolutionary biology of taking geometry seriously in this way. We begin by re-examining the notion of operational point-homology adopted in Chapter 1. In Section 5.1 we will relax its constraints slightly, at some cost in future computer programming effort, to handle the most common circumstance exceptional to the preferred model. In Section 5.2 we consider **allometric coherence**, a new term for patterns of ontogenetic constraint as we identify them from covariance matrices of metric traits, and **allometric reassociation**, the study of phylogenetic changes in these matrices. From comparisons of ontogenies across taxa we arrive in Section 5.3 at new interpretations of heterochrony, paedomorphosis, neoteny, and other classic evolutionary processes. In Section 5.4 we show how the notion of **morphological space**, ordinarily interpreted by reference to characters and their principal components, may be realized more concisely in terms of transformation factors. In a concluding remark we suggest that our readers henceforth think about "size" and "shape" wholly in the context of size change and shape change as they go about biometric research.

5.1 Nonstandard Homology Functions

This section describes certain special situations in which our standard notion of a homology map is ambiguous. The ambiguity can often be exploited to biometric advantage, although we do not yet have the computer algorithms to do so systematically.

5.1.1 Duplication of homologues

A proper morphometric data base consists of sets of landmarks or outline arcs together with a homology map relating each pair of forms. The transformations that are our objects of study, whether expressed by grids or by factors, are intended to depict the homology. But homology is not a unitary concept. Two homology functions according with different principles of correspondence lead to separate morphometric analyses of the same coordinate data. One analysis may be more reasonable than the other, or they may be equally persuasive (or equally flawed). The resolution of conflicts in homology, if resolution exists, lies mostly outside the arena of morphometrics.

One typical ambiguity of homology is associated with growth by accretion. Consider a fish scale, or recall the opercles of Section 4.1.2. In Figure 5.1.1a we note within any single form the record of shapes at a sequence of earlier ages. The point P_1

5.1 Nonstandard Homology Functions

is homologous to point P_3 by virtue of being its ancestor;† yet the actual substance at P_1 earlier is the same material as that at P_2 later, so that this pair ought to be considered to correspond as well. Moss and colleagues (Moss et al., 1980; Skalak et al., 1982) have specified an algebraic framework which accepts both these kinds of ontogenetic correspondence, **anatomical** and **material** homologies (there called "point-paths"). The relation between these two quantifies the control mechanisms by which the bone manages its change of form.

Of the two homology assignments, the material homology is not a proper mapping function between ontogenetic states. New points, such as P_3 in the later image of Figure 5.1.1a, have no material homologue in the earlier form. There is another style of flawed homology in which only a few points have two homologues. When structures (A, B) which abut in one form are separated in another, as in Figure 5.1.1b, there are two natural homologues P_2, P_3 for the point of abutment P_1, one according to homology within organ A, the other according to organ B. When one structure continually slides over another, there is a continual chain of ambiguities generated in this way, just as the Hawaiian Islands represent ambiguity of "homology" with respect to the underlying source of magma. Serial structures show a similar ambiguity: is the third of three fin rays homologous to the third or to the fifth of five rays? There is no general answer. Other difficult cases involve structures that sometimes fuse, as in Figure 5.1.1c, wherein P_1 may be homologous to either P_2 or P_3, or are occasionally absent, as in Figure 5.1.1d.

5.1.2 A hybrid morphometrics for nonsmoothness

Of the several candidates for a biological homology function presented in these examples, none are geometrically smooth maps. Instead, certain points close together in one form are widely separated in another; or certain landmarks pass from inside to outside of substructures; or certain regions in some forms have no counterpart in others.

Other failures of smoothness of homology are associated with the two-dimensionality of most of our coordinate data sets. All the models underlying our examples throughout this book—the medial axis, the biorthogonal grid, the truss—have straightforward generalizations to three-dimensional data. Many spectacular failures of computed homology are associated with the irreducibility of the third coordinate in certain practical applications, for instance in studying ontogeny of the flounder's eye. For other problems, like the study of sharply curving surfaces, no single photograph is sufficient, but a stereopair (or even a pair at 90°) will suffice to generate a data base of accurate distance measures. The transformations in Figures 5.1.2 or 5.1.3 would represent no irregularity of homology if the four points P were midsagittal

---

† More precisely, the osteoblast responsible for the accretion of bone at P_1 is likely the direct ancestor of that responsible for accretion later at P_3.

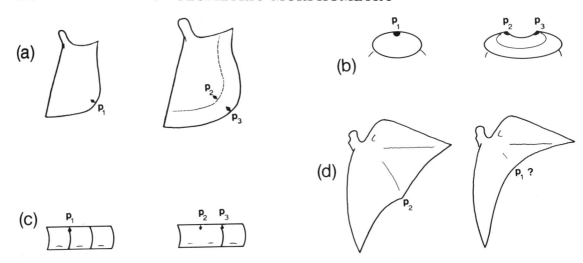

Figure 5.1.1 Ambiguities of homology. (a) In structures that grow by accretion, such as the opercle sketched here, one must choose between anatomical and material homology. There is a sound argument for considering either P_2 or P_3 the homologue of P_1 (see text). (b) When structures separate. (c) When structures fuse. (d) When structures are absent. Should P_1 be assigned a "next best" homologue upon the right-hand boundary?

and Q a bilateral pair. Such a homology function, perfectly smooth in three dimensions, loses its smoothness in any mediolateral projection.

The algebraic and geometrical tractability of the general smooth deformation (see Sec. 4.5) is worth preserving as far as we are able, even in the face of severe technical difficulties. When smoothness of homology is obviously violated, or when homology is ambiguous, we recommend searching for a subset of points which, considered in isolation, sustains the necessary assumptions. If general size is a dominant factor in an analysis, for instance, one might select a subset all of whose interpoint distances load sufficiently strongly on that factor. The other, "nonconforming" landmarks might be considered to have been displaced after that smooth mapping from the positions they should otherwise have taken. Their displacements do not contribute to the computation or characterization of the map, but instead result in separate descriptors; these are vectors with two components, rather than symmetric tensors having three.

For instance, in Figure 5.1.2a we can use the four landmarks P_i to define a smooth mapping of quadrilaterals, and then code the displacement of Q from its predicted position (at the ●) under this mapping. In this example the nonsmoothness is not very marked, so that one might also fit a smooth map in which Q corresponds to its homologue directly (Fig. 5.1.2b). For more extreme failures of smoothness, such as the homology implied in Figure 5.1.3, this latter alternative is no longer available—in the attempt to adjust for the nonconformity at Q the forcibly smooth map is folded, so that part of the interior is pushed outside the boundary.

5.1 Nonstandard Homology Functions

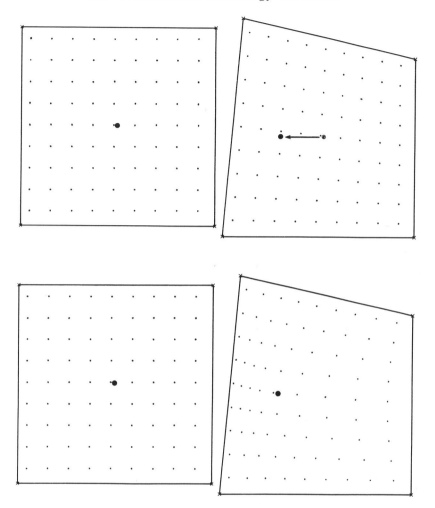

Figure 5.1.2 Two modes for dealing with nonconforming landmarks. (a) Internal landmark related to its imputed position by a displacement vector. (b) Internal landmark deforms the imputed mapping elastically.

As a consequence, a set of landmarks that only roughly satisfies the requirement of smoothness will support many morphometric analyses, one for each subsample of landmarks that can be held to be transforming smoothly. The landmarks omitted from the specification of the homology map become "interior" points, scored each by a transformation vector superposed over the smooth deformation. Semiautomatic selection of a few landmark subsets that provide the "smoothest" analyses, according to some criterion, will perhaps be the fundamental computational step in the future practice of landmark selection.

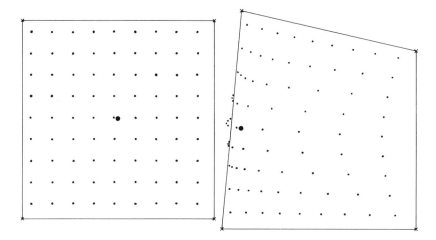

Figure 5.1.3 Limits of the current deformation algorithm. When internal landmarks are displaced by too great a distance in the direction of the boundary, the elastically interpolated mapping is folded over the boundary.

5.2 Allometric Coherence and Reassociation

The publication in 1977 of Stephen Jay Gould's *Ontogeny and Phylogeny* signaled a renascence for the serious evolutionary study of growth and form. Gould argued that the classic terminology of differences in developmental timing among closely related species could be integrated under a single consistent qualitative model, the "clock." Subsequent work (Alberch et al., 1979) expanded these observations into a semiquantitative model involving "onset," "termination," and "rate" of growth— explanatory notions that can sometimes be identified with particular observed measures in data sets carefully selected for this purpose.

We share Gould's goal: a general framework for the simultaneous explication of ontogeny and phylogeny. That discussion would be aided greatly by a general system of measurement such as we have erected in this volume for the particular subject-matter of morphometrics. In the scheme we sketch below, as in Gould's, changes in ontogeny are decomposed into three pure types of change already recognized in the descriptive literature. We differ from Gould and others in preferring measures that are explicitly generated according to our particular class of metric models, so that the quantities explained are variances and covariances of optimally specified lengths. In this way the explanations arrived at are not necessarily limited to the particular set of variables measured. Because factor loadings provide an estimate of dilatations, the findings relate to any morphometric measure, whether or not previously computed, on the available landmarks. We have discussed the connection between deformations and factors elsewhere in this volume (Secs. 1.2, 4.6).

5.2 Allometric Coherence and Reassociation

We have found two modes of application for this framework. Where the study of ontogeny is within species, we call the general approach **allometric coherence**; where between, **allometric reassociation**.

5.2.1 Allometric coherence

The path-analytic factor model of Sewall Wright (Sec. 4.2.3) summarizes the covariance structure of any morphometric data set by latent variables (unmeasured factors) of three kinds (Fig. 5.2.1). The **primary size factor** accounts for as much as possible of the net covariance (exclusive of sequestered blocks of variables) by products of a single set of path coefficients, the primary size allometries. **Secondary factors** account for the blocks of covariances intentionally left unexplained by primary size as construed in this way; their discovery affects the computation of primary size. Depending on context, secondary factors may be shape *discriminators* between populations, *artifacts* of preservation or digitizing protocol, or *functional associations* or other expressions of underlying genetic or epigenetic control. Finally, each measure is assigned a **unique variance** separate from all covariance with the other measures of the model.

We use the term **allometric coherence** to describe the joint application of all three notions in explaining the covariances of a suitable collection of measured variables. The word "coherence" is meant to evoke pleasant speculations on the "unreasonable effectiveness" (Wigner, 1967) of latent variable analysis in biometrical studies. Of the three fragments in this decomposition of covariance, two correspond to classic notions in the geometric study of ontogeny: Huxley's **growth gradient** and Olson and Miller's **morphological integration**.

5.2.1.1 Growth gradients.—The pattern of loadings on primary size† corresponds closely to Julian Huxley's notion of the growth-gradient (Huxley, 1932). As many writers have noted (e.g., Gould, 1966; Bookstein, 1978a), Huxley addressed mainly the pattern of allometric coefficients along a single body axis. The study of rates at a point in all directions (Sec. 2.1) is a geometrically appropriate extension of Huxley's purpose, providing a loading on primary size for every measureable length. Because these loadings are computable from the principal dilatations of a biorthogonal analysis, the field of biorthogonal crosses is the most efficient, least redundant means of transcribing a growth-gradient.

---

† Huxley's allometric coefficients were derived from regressions on size variables explicitly measured. We prefer that this regressor be replaced by a latent variable, that is, a size factor. Furthermore, the particular latent variable ought to be *primary* size, rather than general size, because coefficients of size allometry should take into account the existence of shape or other secondary factors.

$$X_j = a_j Sz + b_{j1} F_1 + \ldots + b_{ji} F_i + e_j$$

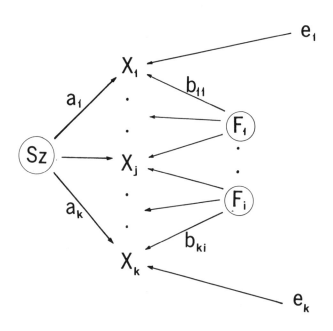

Figure 5.2.1 Path-analytic factor model of allometric coherence. $X_1 \ldots X_k$ are observed log-transformed variables; $a_1 \ldots a_k$ are their loadings on primary size, Sz; $b_{i1} \ldots b_{ik}$ are their loadings on the ith secondary factor, F_i; the error terms $e_1 \ldots e_k$ represent their unique variances.

Huxley objected to Thompson's method of deformation on the ground that it could not take growth changes properly into account in the course of phylogenetic comparisons. This objection is met by the models in Sections 4.6 or 5.2.2, which express both sorts of comparison at once.

5.2.1.2 Morphological integration.—Olson and Miller's *Morphological Integration* (1958) is a milestone in twentieth-century biometry. Its intention was the same as ours: to unite functional and descriptive aspects in the explanation of "interdependence" (covariance) of metric traits over ontogeny. Unfortunately, Olson and Miller implemented this purpose in terms of correlations (that is, coefficients for variables in *pairs*) rather than computing factor loadings for individual variables. They search for **ρF-sets**, sets of highly correlated variables (ρ for correlation) corresponding to F-sets (F for function), sets of variables for which explanations of high correlation are available in advance. Though analysis of correlations in this way is identical with the

5.2 Allometric Coherence and Reassociation

machinery of Wright's path analysis, in Olson and Miller's implementation the data analysis is model-free. They offer us neither parameters to estimate nor numerical values to expect on particular hypotheses. Recently Cheverud (1982) has reminded a new generation of this effort, extending it in interesting ways.

We believe that investigation of morphological integration in this way is subsumed in the search for secondary factors in the general path model. Both methods can search for blocks of variables whose correlations among themselves are larger than correlations with variables outside the set. Such blocks are "explained" by joint dependence on some causal antecedent not shared with the other variables; that antecedent is estimated, recursively, as a linear combination of the variables it accounts for. When causation is represented thus by regression coefficients on latent variables, there results the partition of covariances that path analysis pursues numerically (Sec. 4.2.3). These particular joint causes are sometimes purely geometric, as in Figure 4.2.10; but usually they hint at biologically meaningful associations. For instance, Wright's Leghorn analysis can be considered a study of morphological integration in a set of six variables. In effect Wright finds three F-sets that are also highly correlated—the two head measures, the two wing measures, and the two leg measures—and rejects three other F-sets—the two proximal limb measures, the two distal limb measures, and the set of all four limb measures. (Compare Van Valen, 1965, who finds a ρF-set for the four limb measures; or Pimentel, 1979, in which five distinct pairs of functional associations are identified in principal components.)

Interpreted in a path analysis, Olson and Miller's ρF-sets are secondary factors that have loadings all of the same sign. Clearly, then, factors go unrepresented in ρF-sets whenever their loadings are inconstant in sign. Morphological integration sensu Olson and Miller thus detects only half the biometrical signal. The secondary factor structure admits of both joint causes and compensations, but ρF-sets are invoked only by joint causes. They would not, for instance, recognize the consequences of a gene which turns one growth process on, another off; nor would they detect regulation of the length of parts to the length of a whole.

As noted by Hopkins (1966), in truly longitudinal data there are available what might in this context be called "T-sets," variables concordant in their developmental timing. When a subset of two or more allometric coefficients depends on time in the same way—its traces rising, falling, concave upward or downward, etc.—the variables associated with them may be considered to share a secondary factor that formally accounts for their synchrony. (A variable whose temporal pattern of allometry is unique will instead be assigned unique variance.) Likewise, F-sets of variables that have similar values on Gould's parameters of onset, termination, and rate (as, for instance, endocrinally mediated long-bone measures sharing the timing of the "growth spurt" in humans) will correspond to secondary factors. In the presence of data sampled with sufficient resolution in both space and time, perhaps the three parameters of Gould and others could be extracted directly from a covariance structure in this way. Although these parameters are invoked to explain experimental or comparative findings, their existence is not derived from the raw data by any objective computation.

5.2.1.3 Unique variance.—The third term in the general path model is the **unique variance**, composite of measurement error and developmental error for the variables one by one. By definition this variance is uncorrelated with every other score in the model. In one psychometric version of factor analysis, Guttman's image analysis (1953), the unique variance is computed in advance as the variance unexplained after the regression of any variable upon all the others. (Unfortunately, the scores that result are not uncorrelated with one another.)

In the study of ontogeny, unique variance is to be identified with uncanalized natural variation and the myriad of its unsystematic sources. Its magnitude (in the presence of enough other variables that it may be estimated properly) is a meaningful measure of regulatory precision.

5.2.2 Allometric reassociation

Our thoughts about the importance of shifting covariance structures during the formation of allometrically coherent systems derive from Gould's (1977) arguments on **dissociation**. He discussed the evolutionary consequences of decoupling (dissociating) ancestral associations of ontogenetic "factors" such as "size" and "shape," or developmental rates and timings. Once dissociated, the "factors" may form new associations, that is, new morphologies and ontogenies. We extend the work of Gould and others by imposing the framework of our morphometric models—factor and deformation—upon the phylogenetic study of allometric coherence. For this phenomenon we use the term **allometric reassociation** rather than "dissociation" because the changes involve more than decoupling. The covariance relationships may become stronger (i.e., factor loadings become larger) at the expense of unique variance, which is uncoupled by definition.

In order to work within the framework of ontogeny and phylogeny, the study of allometric reassociation requires comparisons of ontogenies in the context of presumed genealogies. The change in covariance structure from that due to common ancestry is determined by comparing ingroups with outgroups (sensu Hennig, 1966). Even though we are thereby freed from the search for ancestors among recent taxa, our evolutionary conclusions are explicitly tied to the assumed phylogenetic reference system; different genealogical hypotheses will lead to different conclusions (Fink, 1982). In the following discussions, comparisons will be assumed to be in a proper phylogenetic context.

5.2.2.1 Path analysis of reassociation.—In this context, allometric reassociation involves three sorts of changes: changes in growth gradients; changes in secondary size, shape or timing factors; and changes in regulatory precision. Each factor of an allometrically coherent system (Fig. 5.2.1) provides us an explanation for the covariances of the variables with each other and with the factor. If, in a fixed character

5.2 Allometric Coherence and Reassociation

space, the total amount of variation† has remained constant for each of a group of species, then allometric reassociation describes new covariances that have formed "at the expense of" relatively primitive patterns of covariance or unique variance. That is, the same variance has been partitioned in different ways. New covariance relationships are manifested as novel secondary factors, or as larger or smaller loadings for previously existing factors. If the total variation is conserved, then any increase in covariance implies increased canalization and less developmental noise. In this case, allometric reassociation describes the change in shape of the occupied hypervolume in character space.

Alternatively, the change in allometric coherence may incorporate a change in the total amount of variation. When novel, stronger, or weaker covariances are formed, it is not necessarily at the expense of previously existing relationships. Increased canalization is detected from an increase in loadings; an increase in covariances alone may indicate only an increase in the range of one factor (e.g., primary size). The occupied hypervolume may have changed its "size"—that is, its trace—as well as its shape.

We may apply the factor models of allometric coherence and reassociation to issues in evolutionary biology, such as those concerning macroevolution. Macroevolution is defined as sudden "large-scale" evolutionary change that may result in "higher taxa" above the level of species (e.g., Stanley, 1979). We shall not discuss the many serious problems with macroevolutionary studies (but see Bookstein et al., 1978; Schopf, 1982), including the definition of "large-scale" morphological change; however, if supposed macroevolutionary changes correspond to the appearance of new factors or factor structures, then the partition of covariance may lead to a general description of these evolutionary events, independent of the magnitude of the morphological change. In similar fashion, metric descriptions of other types of evolutionary change may be obtained in terms of allometric reassociation of covariances.

5.2.2.2 Biorthogonal analysis of reassociation.—In Section 1.2 we noted that factors and deformations are two aspects of the same model. Hence, regularities of covariance change are interpretable in terms of dilatations, orientations of biorthogonal crosses, and complexity. In path analysis, estimates of coherence and reassociation depend upon the measurement scheme. The same estimates when derived from biorthogonal analysis depend upon the system of landmarks and boundary curves. The amount of localized information derived from the deformations can be controlled by adjusting the density of landmarks at which the homology function is observed.

Allometric coherence incorporates primary size and secondary factors plus unique variance in the description of an ontogeny. A biorthogonal description for each

---

† In this context, that amount is not a "generalized variance" but the sum of the variances of the separate characters—the trace of their covariance matrix, as partitioned in the course of principal components analysis.

of the factors can be calculated, given that the factors are known. Because only two forms are compared in biorthogonal analysis, there is not sufficient information about variation for any deformation to be informative about more than one factor. The change in form due to any factor may be analyzed by controlling for the other factors. For example, primary size is inherent in the dilatations and principal directions of change between the average small and large forms. The use of average forms removes the effects of secondary factors and unique variance. To calculate the biorthogonal description of a secondary factor, one compares the average forms at the morphological extremes of the factor at the same primary size.†

Estimates of allometric coherence may be obtained locally (for each locus), regionally, and across the entire form. Regional and global estimates require calculation of the complexity of the ontogenetic transformation. This heterogeneity is fundamental to notions of allometric coherence because it measures the independence of developmental trends among regions of the form.

Allometric reassociation involves three sorts of changes: changes in the gradients of dilatations; changes in the principal directions of form change; and changes in the profiles of regional complexity (e.g., Fig. 4.7.9). We may use these changes to test hypotheses about the evolution of form. Gould (1977) speculated that decoupling of ancestral developmental factors might result in shifts to new morphological plateaus. New patterns of principal directions and higher regional complexities in derived taxa indicate that relatively primitive developmental trends have been decoupled, resulting in new, divergent developmental trajectories. Alternatively, if a derived taxon retains the primitive directions of form-change but has lower regional complexities, then increased canalization of the primitive developmental trend has occurred. Whether decoupling, canalization or some combination thereof, allometric reassociation is the model in which to examine morphology in an evolutionary context. Generalizations about prevalent mechanisms should be derived from studies of reassociation across a variety of clades. An advantage of using biorthogonal analysis for this purpose is that the descriptors of form-change—dilatations, principal directions, and complexity—provide a common currency for comparing results among studies.

5.3 Developmental Programs and Evolutionary Inference

Geometric morphometry and evolutionary process have in common the language of change. The former is the geometry of form-change, while the latter is the mechanism of genome-change. Current thinking in evolutionary biology (Gould, 1977; Alberch et al., 1979; Fink, 1982) suggests that particular developmental modifications reliably correspond to particular evolutionary processes. We speculate that geometric criteria can be used to choose among alternative evolutionary processes.

† At present, we can change the primary size of an archive of landmarks only with a truss network (Sec. 4.4).

5.3 Developmental Programs and Evolutionary Inference

5.3.1 Survey of conventional terminology

The link between geometric morphometry and evolutionary process is the developmental program. The sequence of stages through which an organism passes in its lifetime is partly under genetic control. This genetic control is here referred to as the **developmental program**. Developmental programs (along with epigenetic phenomena) cause changes in shape and size during life-cycles and we may quantify the geometry of these form-changes using biorthogonal analysis. Each evolutionary process engenders a pattern of form-change, in terms of the complexity and amount of change (see Sec. 4.7) observable among ontogenies of related species. The logical basis for these comparisons is a phylogenetic hypothesis of species relationships.

A phylogenetic context is necessary to prescribe a choice of comparisons among ontogenies (Eldredge and Cracraft, 1980; Fink, 1982). This is because inferences about evolutionary processes depend upon the hypothesized evolutionary sequence among members of the group being studied. For example, assume that taxon B is a descendant of taxon A, and that for some morphological feature taxon B possesses a new stage appended to the last of its ancestor. However, were we to take taxon A as the descendant, we would conclude that it evolved by deletion of some stages present in the ancestor. Without a well-corroborated phylogenetic hypothesis, specification of a particular evolutionary process is impossible. (Presumably the phylogenetic hypothesis is not confounded by the types of processes that we are trying to infer.)

Among the developmental-evolutionary processes, the class most compatible with geometric analysis is heterochrony. **Heterochrony** refers to changes in relative timing of appearance and rate of development for characters already present in ancestors (Gould, 1977). It has two conventional subdivisions. **Paedomorphosis** accounts for species in which the terminal stages resemble pre-terminal (earlier) stages of their ancestors—developmental processes have been truncated or slowed with respect to more primitive species. **Peramorphosis** (Alberch et al., 1979) accounts for species in which adult stages incorporate new morphologies by acceleration of developmental rates or by delay in the termination of development.

Note that these rates are with respect to denominators that are unspecified. These may be calendar time, or generation time, or the timing of a developmental marker. Different inferences may result from choices of different developmental markers or measures of time; this problem is beyond the scope of our text.

We will outline the application of geometric morphometry to evolutionary inference with the aid of Fink's (1982) example (Fig. 5.3.1a). We assume three taxa: an outgroup T_o, and two sister taxa T_{paed} and T_{per}. In discussion of the evolutionary processes that follow, taxon T_{paed} will be taken as paedomorphic, and taxon T_{per} as peramorphic with respect to taxon T_o. There are three classically recognized varieties of paedomorphosis: progenesis, neoteny and post-displacement. These correspond to the three different ways of being *below* the diagonal in Figure 5.3.1b–d. There are also three named varieties of peramorphosis—hypermorphosis, acceleration and pre-displacement, the three different ways of being *above* the diagonal. The six processes

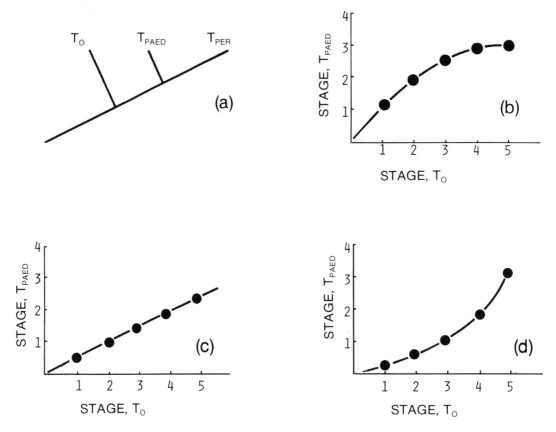

Figure 5.3.1 Diagrams for the discussion of heterochrony. If both taxa were developing at the same rate, the plotted curves would all lie along the 45° diagonals. (a) Cladogram for three taxa, after Fink (1982). T_o, outgroup; T_{paed}, paedomorphic descendent; T_{per}, peramorphic descendent. (b) Relation of stages when T_{paed} is derived from T_o by progenesis. (c) Same, by neoteny. (d) Same, by post-displacement.

appear to correspond to perturbations, positive or negative, in the three basic parameters given by Alberch et al. (1979).

5.3.2 Format for quantitative investigation

The terminology of heterochrony is somewhat abstract. As Fink (1982:254–255) notes, "[Gould and Alberch et al.] do not provide explicit descriptions of the procedure for detecting those results in nature. ... Heterochrony is discussed by these authors in the context of 'ancestral' ontogenies, but it is obvious that for the overwhelming majority of species, ancestors are not specifiable, much less their

5.3 Developmental Programs and Evolutionary Inference

ontogenies. In the absence of direct observation of an ancestor, how can the results of heterochronic developmental processes be recognized?" He suggests the consideration of quantitative, unspecified shape traits (called σ, as in Alberch et al.). The necessary comparisons of ontogeny are executed by subtraction of the values of σ at appropriate pairs of developmental times, or comparisons of the slopes of σ as a function of the developmental time.

But we can make this procedure a great deal more operational. The shape contrasts which are the objects of consideration are precisely the quantities computed by the biorthogonal method. This geometric computation handles variations in shape-change both by age and by region of the form, as we exemplified in Section 4.7. We remain free to choose as our index of "developmental time" a criterion of age, size or stage as appropriate.

To proceed with an investigation, we must first decide whether or not heterochrony describes a particular set of ontogenies. Heterochrony presumes shared early developmental sequences between organisms and their outgroups before divergence (Gould, 1977; and see Fig. 5.3.1). These shared sequences may be verified from the amount, complexity or directions of shape-change. If an organism develops through the same series of shapes as its outgroup, then ideally, the pattern of regional complexities or directions of shape-change over this series will be identical for each ontogeny. It may be easier in practice to search for comparisons across ontogenies that have zero amount of shape-change.

Consider, for example, the case of progenesis (Fig. 5.3.1b). Here taxon T_{paed} develops as its outgroup T_o for some period of time, after which its development is truncated with respect to that of T_o. Without having any specific shape measure σ in mind, we can nevertheless delineate expectations about our principal measures of shape dissimilarity. In effect we are measuring vertical distance in the plots of Figure 5.3.1 without any need for a vertical axis, as long as we are confident about the shared sequences. For progenesis, we expect, among other features, the following:

1. The mean amount of shape change between successive ages within taxon T_{paed}, which represents the slope of the regression on time of the optimally changing shape measure, will drop toward zero earlier than the corresponding index within taxon T_o.

2. The mean amount of shape-difference between corresponding ages of T_{paed} and T_o will increase abruptly beyond the age of truncation (shape stage 3 of the outgroup).

3. The mean amount of shape-difference between forms of T_{paed} with age-lagged forms of T_o (i.e., forms of T_o corresponding to earlier ages in T_{paed}) will decrease abruptly just after the age of truncation.

This circumstance, progenesis, can be discriminated from a mere plesiomorphy by considering the ontogeny of T_{paed} in relation to those of T_o and T_{per}, as indicated by

5 GEOMETRIC MORPHOMETRY

Fink (1982). Comparisons of outgroups and ingroups provide the basis for deciding where on the cladogram the observed difference(s) among ontogenies occurs. We employed such a three-taxon comparison to analyze evolutionary shape-change within the Molidae (Sec. 4.5.3.1). Similar criteria may be delineated for discerning the other five varieties.

5.3.3 Without a time scale

The preceding analysis explicitly demands a prior notion of "corresponding stages," a shared time-scale. In the ordinary practice of evolutionary biology this correspondence of time scales is usually arbitrary. We have no basis for comparison of biological clocks, other than the morphologies we are studying in the context of the clocks. Should we wish not to confound the object of study, which is change of ontogeny, we can still distinguish many categories of change in ontogenetic program.

Assume that, over some region(s) of the form, two ontogenies, of primitive taxon and derived taxon, have their principal axes in common. In the vicinity of these regions each ontogeny can be described by a sequence of biorthogonal crosses, each with dilatations d_1, d_2, representing the deformation of ontogenetically successive forms from an arbitrary early stage. The dilatations for a single taxon can be plotted as a curve in the (log d_1, log d_2)-plane, then rotated 45° to the plane of size-change vs. shape-change, log d_2/d_1 vs. log $d_2 d_1$ (Fig. 5.3.2a; compare Fig. 4.7.3). The curve bears an implicit time scale which serves as an index of the second form in each comparison. But even if we have not measured age, we still have the autonomous curve (Fig. 5.3.2b) showing the dilatations with respect to each other without explicit reference to time. Because we assumed consistency of alignment of all grid directions, selection of a different age of reference is equivalent to translations of this curve. The shape of the curve is unaffected by translation.

Specifically, its slope is independent of that choice of starting form. This slope,

$$\frac{\log d_2 - \log d_1}{\log d_2 + \log d_1}$$

is "rate of change of shape-change with respect to size-change"; it is independent of choice of a time scale. Likewise the contrast of two of these slopes is independent of time scale. Were the time scale fixed, we could determine whether an increase in slope from one line to another expressed a speeding-up of development of d_2 or instead a slowing-down in d_1 (Fig. 5.3.2c). But because we are free to vary the time scale, we cannot tell either of these two possibilities from its opposite—in the context of a "time" scaled as d_2, d_1 has slowed; if, on the other hand, "time" goes as d_1, then d_2 has slowed.

For any pair of dilatations along principal directions, and for any choice of two stages each from a pair of ontogenies, we say that the descendant species shows **shape**

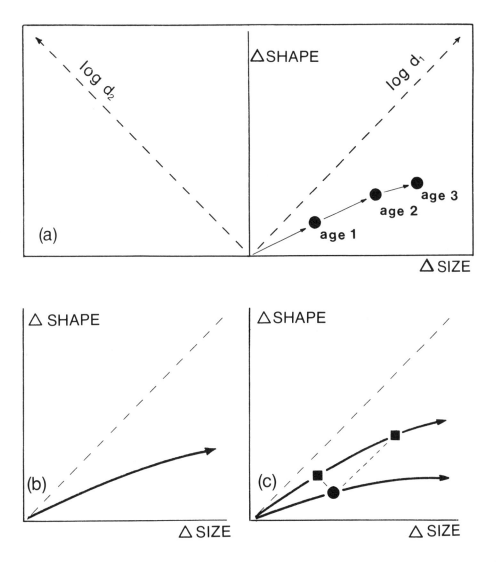

Figure 5.3.2 The curve of shape-change as a function of size-change. (a) Time scale known. (b) Time scale omitted; the slope of the curve is still well-defined. (c) Determining which of d_1, d_2 is responsible for a change in slope requires knowledge of an exogenous time scale. Without it, we cannot determine which stage of the primitive ontogeny (squares) is appropriate for comparison with a stage from a derived ontogeny (circle). Left-hand square, time scaled as log d_1, and the derived form has lower log d_2: that is, log d_2 has slowed. Right-hand square, time scaled as log d_2, and the derived form has lower log d_1: that is, log d_1 has slowed. But these two interpretations are contradictory.

acceleration, **shape deceleration**, or **shape isomery** with respect to the size of the ancestor, as the slope of its curve is greater than, less than, or equal to the slope of the ancestral curve. This comparison between two ontogenies is independent of ontogenetic time scale but depends, of course, on the choice of pairs of stages from the two ontogenies.

It is now appropriate to restore a little bit of information about biological time to the analysis. We are usually able to find external temporal markers, such as reproductive maturity or a discontinuity in the change of dilatations elsewhere; or we may notice a pair of relatively young forms from sister taxa that have approximately the same shape, or the same size and shape. In all these cases we can superimpose the curves at that pair of points.

From such a shared moment of reference, these autonomous ontogenetic curves may be traced forward and backward in ontogenetic time. They may lie atop each other (Fig. 5.3.3b) or may diverge linearly or nonlinearly in various ways (Fig. 5.3.3a,c). Naturally any pair of dilatations along temporally stable principal directions may be diagrammed in this way, whether or not they refer to the same biological locus. The resulting framework for description of alterations in ontogenetic rates is very rich indeed.

The multivariate version of this point of view, treated by Hopkins (1966), assigns factors to particular profiles of allometry with respect to general size: factors for allometric coefficient constant over time, rising, falling, curving upward, curving downward, and so on. The scaling of time is irrelevant in this context also.

With such tools one can address important questions in evolutionary biology. For instance:

1. What are the major alterations among developmental programs for size- and shape-change that have occurred during the evolution of a group?

2. Are the ontogenetic sequences of shape-change of outgroups retained in the ontogenies of descendant species?

3. Are intermediate developmental stages intermediate in shape?

4. How much form-change has occurred during the evolution of a group of organisms?

Answers to questions such as these can be interpreted in the context of the following paradigms: (1) that development proceeds from the more general to the less general, (2) that developmental stages of more primitive taxa are retained in the ontogenies of less primitive relatives, and (3) that differences in rates and timing of development describe the formation of new ontogenies. Studies along these lines could be carried out in an experimental and explicitly geometric fashion.

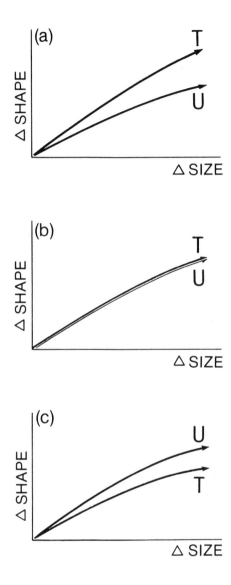

Figure 5.3.3 Rates of change of shape-change with respect to size-change. (a) *T* shows shape acceleration with respect to *U*. (b) Shape isomery. (c) *T* shows shape deceleration with respect to *U*.

5.4 The Investigation of Morphological Spaces

One cannot conveniently determine why certain morphologies do not exist until one can visualize them. This idea suggests the consideration of **morphological space** (or **morphospace**), the arena for comparison of real forms to other imaginable forms (Lull and Gray, 1949). Raup's (1966) geometric analysis of shell coiling in molluscs introduced methods for studying unobserved forms by reconstruction. In this section we first consider a geometric model of evolution in morphospace leading to certain expectations about the visual appearance of branching sequences. We then describe how the truss protocol permits us to reconstruct, in morphospace, forms more complex than those of mollusc shells. In this way we may study the limits of occupiable regions in morphospace to better understand observed distributions of taxonomic or ecological assemblages.

5.4.1 Divergent evolution and taxonomic dissimilarity

The phylogeny of a monophyletic group of organisms may be modeled as a series of branching events in morphological space. If directions of branching are morphologically unrelated, then the "planes" of pairs of branches will be randomly oriented with respect to one another. Because we mean morphospace to encompass all aspects of form, its dimensionality is too high to be estimated. We therefore know nothing about the geometry of the phylogenetic tree in this space except that there were original units of distance along its branches, such as years, DNA substitutions, etc. Measuring a sample of descriptive characters is equivalent to projecting the tree onto a hypersurface, a subset of the full space, spanned by the subset of variables. In this curving subspace we have distorted or lost the original unit of the geometry. As a proxy for evolutionary time we must compute some other measure of distance between forms.

The modeling of phylogenetic trees in morphospace unites two aspects of evolution, shape-change and genealogical distance, into a single concept measured in alternate ways. This general concept is "taxonomic dissimilarity" which, like body size, cannot be defined explicitly enough to be measured directly. It is instead another latent variable, which accounts for most of the associations observed among a number of possible indicators. Any particular measure of dissimilarity ought to correlate (however imperfectly) with other measures of dissimilarity, such as (1) analogous measures of distance using any other arbitrary subspace of variables, (2) the number of branches or nodes along the shortest path connecting two taxa, (3) the traditional assignment of taxonomic distance by lowest common rank (e.g., subspecies, genus, etc.), and (4) relative position on a genealogical tree.

5.4 The Investigation of Morphological Spaces

From two measures of dissimilarity we can compute only a single correlation.† With three variables we can quantify their correlations with the latent factor they ostensibly share. From four or more variables we can discover secondary factors explaining some of the information not included in the best pooled factor (Sec. 4.2). These variables are all equivalently arbitrary estimates of the same phenomenon: extent of evolutionary change over time.

We would expect that almost all composite measures of distance among taxa will correlate with "general taxonomic dissimilarity" (i.e., with each other) to the extent that they do not express particular factors, such as "habitat" or "size." Variation in body size, for instance, violates the model of random multidimensional orientation of branching events because it represents a consistent and biologically interpretable ranking of species along a single dimension. We have arrived at a paradox in the study of morphological evolution: to measure distances along the evolutionary tree most meaningfully, one must measure separate forms using a character set that taps each factor no more than once. To the extent that there exist biologically interpretable (e.g., allometric) covariances among characters, the basic geometric model of random branching-orientation is inapplicable. Such covariances must first be sequestered; only then will the characters selected engender a meaningful projection of the phylogeny onto the hyperspace delimited by the subset of variables.

From a multitude of projections of morphological space we could, in principle, learn more and more about the geometry of the phylogenetic tree. For this reason the analysis of a single set of 100 descriptive variables would be less informative than ten analyses of ten random subsets of the variables. From ten subsets we would learn something about the variation of distance measures associated with the arbitrariness of the projection. In particular, we might be able to detect failures of the general model of random branching-orientation, failures due perhaps to parallelisms or convergences of form among taxa. Such failures are likely to be observed in some projections and not in others. Using a sufficiently refined set of variables, we might also be able to distinguish regions of the morphospace that could not be occupied by any reasonable biological form.

5.4.2 Examination of unoccupied morphological space

Even if hyperplanes of branching are independent in orientation, observed taxa might not be randomly distributed throughout the morphological space that describes

---

† For this reason it is insufficient to compare differences in body shape only with traditional taxonomic distance among groups (e.g., Douglas and Avise, 1982; Cherry et al., 1983). Classical taxonomy need not provide any better or worse estimate of overall dissimilarity than can be derived from a properly chosen set of variables. Indeed, when one is not considered to be merely an error-prone estimate of the other, the differences between the two estimates ought to be very interesting.

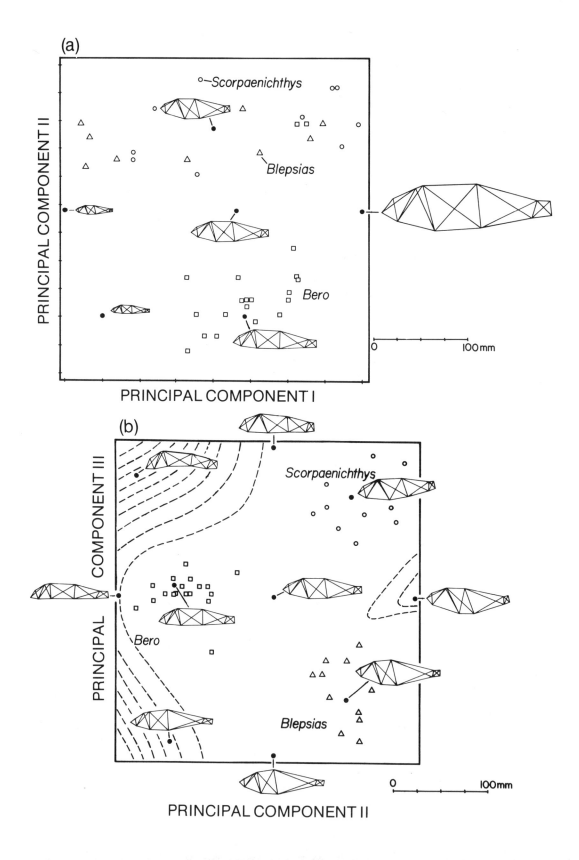

5.4 The Investigation of Morphological Spaces

the total spectrum of theoretically possible forms. That is, if we were able to reconstruct or envision the total morphospace for some groups of organisms, not all of the space would be occupied. Areas without forms might exist for a number of reasons: (1) they may represent mechanical gaps, combinations of characters that result in physically impossible morphological configurations; (2) they may be the result of phylogenetic constraints, where certain combinations of characters may not have been accessible for historical reasons; (3) even if a region of space may have been accessible to a particular lineage, the proper combination of selective forces may not have been present to guide the lineage into the gap; (4) the gap may have been created by extinction; or (5) the gap may be occupied by some form not yet studied. Our view of morphospace is dependent upon the particular subset of variables we choose to study. If our partial description of morphology is reasonable, then in any projection we should be able to distinguish at least two sorts of unoccupied regions: those empty owing to physical constraints, and those unoccupied as a result of historical or sampling accident.

The truss protocol (Sec. 4.4) provides a class of geometrically reasonable sets of variables, the principal components of which approximate factors of covariation. In addition to systematically describing overall body form, the protocol allows the reconstruction of forms predicted at any point in morphospace. Figure 5.4.1 illustrates the procedure, in which the morphospace has been further projected into three dimensions—the first three principal components of a sample of truss measures for three Pacific cottids. The original log-distance measures are linear combinations of scores on the components; thus any point in the space specifies a form by predicting all of its individual character states. Because scores must be assigned to the full set of principal components to uniquely specify a form, the fourth through last components have been sampled at the pooled centroid.

The process of fitting the truss provides a measure of **strain** which quantifies the degree of incompatibility among the individual measurements. Each point in morphospace has an associated strain; the isoclines of Figure 5.4.1 are an arbitrary sampling of the strain field of the first three components. As strain increases away from the observed forms, the predicted truss measures become mutually incompatible. Examination of reconstructed forms, associated component loadings, and strain statistics allows us to understand how forms become internally inconsistent outside of occupied regions of morphospace. By delimiting the full extent of the low-strain

Figure 5.4.1 Example of the reconstruction of body forms at unoccupied points in morphological space. (a) Principal components I and II of samples of truss measures for three Pacific cottid species: *Bero elegans*, *Blepsias cirrhosus*, and *Scorpaenichthys marmoratus*. (b) Principal components II and III for the same samples. Each hollow symbol represents a single measured specimen; solid circles represent positions of the predicted forms shown. Strain isoclines are sampled at intervals of 0.2, beginning at 1.0. Below this value, strain isoclines are highly irregular in contour. Reconstructed forms are scaled in millimeters.

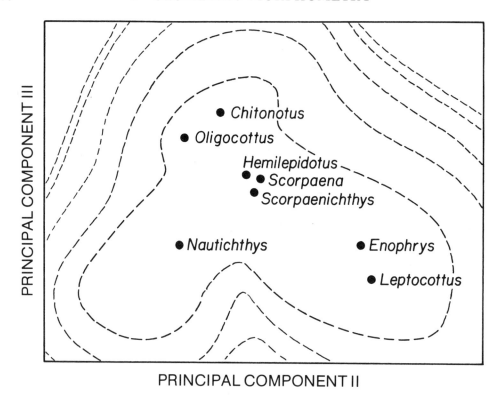

Figure 5.4.2 Principal components II and III of the morphospace of seven genera of Pacific sculpins and the outgroup genus *Scorpaena*. Cottid genera are represented by samples of *Chitonotus pugetensis*, *Enophrys bison*, *Hemilepidotus spinosus*, *Leptocottus armatus*, *Nautichthys oculofasciatus*, *Oligocottus maculosus*, and *Scorpaenichthys marmoratus*. Strain isoclines are sampled at intervals of 0.5, beginning at 1.0.

region, we may interpret observed variation in shape in relation to that which is mechanically possible.

The distribution of taxa in morphospace will corroborate a phylogenetic hypothesis when monophyletic groups are localized in the space. Alternatively, local groups might be caused by functional or ecological convergence. Consider a truss scheme for seven genera of Pacific sculpins and an outgroup, the scorpaenid genus *Scorpaena* (Fig. 5.4.2). As in the previous example, we reduce the usual set of truss variables to the first three principal components and use truss reconstruction to draw the forms. As before, the first component is accounted for mostly by within-group variation in body size. The taxa are distributed in clusters within the space of the second and third components (sampled, in this figure, at the centroid of the first and fourth through last components). The strain field defines a three-lobed region of low strain, one lobe of which extends well beyond the observed taxon (*Nautichthys*).

5.5 Recapitulation and Prospect

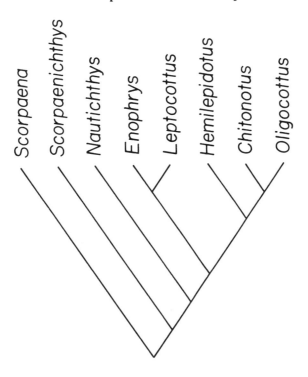

Figure 5.4.3 A cladogram of the eight genera of Figure 5.4.2. Data from Bolin (1947).

When we superimpose on this morphological space an independently derived phylogenetic hypothesis (Fig. 5.4.3), we are able to visualize trends of apparent divergence along different phyletic lines (Fig. 5.4.4). The more primitive forms (including the outgroup) cluster in the central region of the occupied subspace presumably due to primitive similarity. The cladogram diverges into three derived regions, each characterized by a distinctive body form. If we append to this analysis additional information, such as habitat, feeding preferences, etc., we may proceed from a phylogenetic hypothesis to a more detailed examination of the effects of environment on morphological evolution. A similar methodology might be used to investigate resource partitioning and coevolution of fishes or other organisms.

5.5 Recapitulation and Prospect

The dominant theme of systematics is, of course, comparison. Shared by all its schools, that thrust pervades every corner of its logic and methodology—with one ironic exception: the biometrics of form-description. By and large, organisms are measured one at a time; afterwards, only the measurements are compared. In this treatise we extend the comparative theme backwards in the research cycle to the

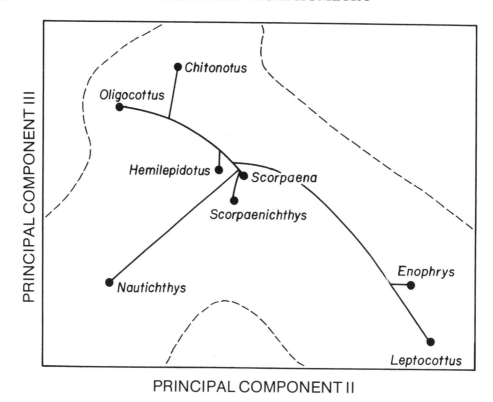

Figure 5.4.4 The topology of the cladogram of Figure 5.4.3, superimposed on the projected morphospace of Figure 5.4.2.

moment when form is first reduced to quantity. In the morphometric discipline we have suggested, geometry and biology play roles as commensurate as possible. Geometric information enters as planar or spatial locations of points and curves; biological information enters by way of the homology map, the function relating the set of geometrical locations describing one form to the set of locations describing any other.

Throughout our text we have contended that morphometric quantifications can and ought to embody both of these sources of information at once. The two channels should not be separated at any time in the subsequent processing and analysis of morphometric data. We argue, for instance, that we cannot effectively measure shape or size before measuring shape-change, which is an explicit comparison; that the classic terminology of tempo and mode in evolution implicitly invokes certain quantifications of homology maps; and that in studying the phylogeny of organisms that grow, we should model both size-change and shape-difference explicitly as homology maps.

We do not consider these different versions of the fundamental tenet to be original with us, but our emphasis upon them is perhaps unusual. When we criticize other methods, such as Fourier analysis of outlines or conventional multivariate

5.5 Recapitulation and Prospect

morphometrics, it is because the quantities that result seem to us to have been purged of biological information. These practices do not necessarily lead to errors in findings, but they are inefficient uses of the available data, discarding at every step information that would add a layer of biological interpretation to the conclusions or coefficients that emerge. The pioneers of a more informative morphometrics—Thompson, Wright, Olson and Miller, Blum, and others—knew better than this; we have retrieved their fundamental suggestions and built upon them.

In this treatise we have attended relatively little to conventional tests of statistical significance. Methods have just been developed for rigorous statistical comparisons of populations of shape changes displayed by the biorthogonal formalism (Appendix A.4). It is convenient to have access to such machinery; but the principal intellectual task is prior to that stage, in the construction of the actual quantities of size and shape comparison used by the statistician for computation of probabilities. Extension of other statistical formalisms—variance, covariance—to sets of these comparisons is presently under development. We do not expect that levels of significance for population differences will be greatly changed from those computed according to other measurement schemes; but the description of the nature of the differences may be greatly altered.

More important is the extension of these new methods to various additional types of data beyond those dealt with in Chapter 3. We have speculated on the handling of nonstandard "landmarks" in Section 5.1. For data in either two dimensions or three, we need protocols for combining descriptors of the medial axis with descriptors of landmark configurations for the relation of internal structures to boundaries. We need formalisms for combining information from homologous landmarks with descriptions of curves by extrema of curvature, inflections, or extremal coordinates in some arbitrary coordinate system.

At present we have no protocols for phenetics, cladistics, or phylogenetics on such data sets as those of Tables 4.7.1 or 4.7.2: dissimilarity matrices whose entries are vectors or two-by-two-by-two tables. We lack experience in describing the evolution of patterns of size transformation or shape polymorphism as expressed in factor loadings; and we lack descriptors of evolutionary lineages based on the morphometrics of successive form-changes (Bookstein, 1980a). Though we acknowledge the magnitude of these gaps, we expect that they will be filled as compelling hypotheses requiring their application are encountered in the course of other studies.

We have emphasized the flexibility of these new biometric quantities, particularly for systematics. Evolutionary biology is not alone in this time of transition to a richer biometrics. There are many other fields that must measure biological form—experimental biology, biomechanics, the clinical medical study of growth and of disease. There the same dilemmas flow from the conventions of measuring forms separately, and the same solution—the direct measurement of form-comparison—resolves problems from orthodontic treatment planning or monitoring of myocardial infarction to embryogenesis of salamander limbs. We believe we face a reorientation of all biometrics, a radical expansion of the effectiveness with which size and shape information are exploited throughout biology.

Appendices

A.1 Some Useful Coordinate Formulas

The following formulas may be useful in statistical analyses of coordinate data. They deal with observed **points** such as $P = [p_1, p_2]$, and with straight lines such as $L = [l_1, l_2, l_3]$. The line $[l_1, l_2, l_3]$ contains those points $[p_1, p_2]$ for which $l_1 p_1 + l_2 p_2 + l_3 = 0$. For instance, the line "$y = 3x + 2$" is written as $[3, -1, 2]$. For any constant $c \neq 0$, the lines $[l_1, l_2, l_3]$ and $[cl_1, cl_2, cl_3]$ are the same, so that of these three parameters only two are independent.

1. The distance between two points $P = [p_1, p_2]$ and $Q = [q_1, q_2]$ is

$$\sqrt{(p_1 - q_1)^2 + (p_2 - q_2)^2} \, .$$

This will be written as D_{PQ} in the formulas following. ("D" here stands for "denominator," not "distance.")

2. The line through $[p_1, p_2]$ and $[q_1, q_2]$ is

$$[p_2 - q_2, \ q_1 - p_1, \ p_1 q_2 - q_1 p_2]$$

3. The point midway between $[p_1, p_2]$ and $[q_1, q_2]$ is

$$\left[\frac{p_1 + q_1}{2}, \ \frac{p_2 + q_2}{2} \right]$$

4. The line parallel to $[l_1, l_2, l_3]$ through $[p_1, p_2]$ is

$$[l_1, \ l_2, \ -(p_1 l_1 + p_2 l_2)] \, .$$

5. The cosine of the angle from one line $L = [l_1, l_2, l_3]$ to another line $M = [m_1, m_2, m_3]$ is

$$\frac{l_1 m_1 + l_2 m_2}{D_L D_M} \, ,$$

where $D_L = \sqrt{l_1^2 + l_2^2}$ and $D_M = \sqrt{m_1^2 + m_2^2}$. The sine of this same angle is

$$\frac{l_2 m_1 - l_1 m_2}{D_L D_M}.$$

6. Starting with points rather than lines: the cosine of the angle at $P = [p_1, p_2]$ between the line to $Q = [q_1, q_2]$ and the line to $R = [r_1, r_2]$ is

$$\frac{(p_1 - q_1)(p_1 - r_1) + (p_2 - q_2)(p_2 - r_2)}{D_{PQ} D_{PR}}$$

and the sine (measured from the line PQ to the line PR) is

$$\frac{(p_1 - q_1)(p_2 - r_2) - (p_2 - q_2)(p_1 - r_1)}{D_{PQ} D_{PR}}$$

where D_{PQ} is the distance from P to Q from formula #1, and likewise D_{PR}.

7. The line perpendicular to $[l_1, l_2, l_3]$ through the point $[p_1, p_2]$ is

$$[-l_2,\ l_1,\ (l_2 p_1 - l_1 p_2)].$$

8. The perpendicular bisector of the segment between $P = [p_1, p_2]$ and $Q = [q_1, q_2]$ is

$$\left[q_1 - p_1,\ q_2 - p_2,\ \frac{q_1^2 + q_2^2 - p_1^2 - p_2^2}{2} \right]$$

Points on this line are the same distance from P as they are from Q.

9. The intersection of two lines $L = [l_1, l_2, l_3]$ and $M = [m_1, m_2, m_3]$ is the point

$$\left[\frac{l_2 m_3 - m_2 l_3}{D_{LM}},\ \frac{l_3 m_1 - l_1 m_3}{D_{LM}} \right],$$

where $D_{LM} = l_1 m_2 - l_2 m_1$. If $D_{LM} = 0$, the lines are parallel or identical.

10. The distance of point $P = [p_1, p_2]$ from line $L = [l_1, l_2, l_3]$ is

$$D_{LP} = \frac{p_1 l_1 + p_2 l_2 + l_3}{D_L}$$

where D_L is as in formula #5. D_{LP} is zero, of course, when point P is on line L.

A.1 Some Useful Coordinate Formulas

11. The reflection of point $P = [p_1, p_2]$ in line $L = [l_1, l_2, l_3]$ is the point

$$\left[p_1 - 2 \frac{D_{LP}}{D_L^2} l_1,\ p_2 - 2 \frac{D_{LP}}{D_L^2} l_2 \right],$$

where D_{LP} is as in formula #10 and D_L is as in formula #5. The foot of the perpendicular from point P to line L is the point

$$\left[p_1 - \frac{D_{LP}}{D_L^2} l_1,\ p_2 - \frac{D_{LP}}{D_L^2} l_2 \right].$$

12. The angle bisectors of the lines $L = [l_1, l_2, l_3]$ and $M = [m_1, m_2, m_3]$ are the lines

$$\left[\frac{l_1}{D_L} + \frac{m_1}{D_M},\ \frac{l_2}{D_L} + \frac{m_2}{D_M},\ \frac{l_3}{D_L} + \frac{m_3}{D_M} \right]$$

and

$$\left[\frac{l_1}{D_L} - \frac{m_1}{D_M},\ \frac{l_2}{D_L} - \frac{m_2}{D_M},\ \frac{l_3}{D_L} - \frac{m_3}{D_M} \right]$$

where D_L, D_M are as in formula #5.

13. The area of the triangle with vertices $[p_1, p_2]$, $[q_1, q_2]$, $[r_1, r_2]$ is

$$\frac{(p_1 q_2 - q_1 p_2) + (q_1 r_2 - r_1 q_2) + (r_1 p_2 - p_1 r_2)}{2}.$$

This quantity is positive when the vertices are listed in counterclockwise order, negative when they are in clockwise order. The area of a polygon with vertices P^1, P^2, ..., P^n is the sum of the areas of the triangles $P^1 Q P^2$, $P^2 Q P^3$, ..., $P^{n-1} Q P^n$, $P^n Q P^1$, where Q is any point whatever.

14. For any l_3, l_4, the line

$$l_\theta = [l_1 \cos \theta + l_2 \sin \theta,\ -l_1 \sin \theta + l_2 \cos \theta,\ l_4]$$

makes an angle θ counterclockwise from the line $[l_1, l_2, l_3]$.

A.2 Biorthogonal Analysis of Triangles by Hand

In Section 2.1 we reviewed the sources of information in biological outlines: landmark locations and the curving arcs between them. Often the curving is systematic or typical over most forms of a data set, as in the atherinids and cottids of Section 4.6. Whenever that is the case, for the purpose of shape comparison one may discard the information the curves bear, relying on landmarks alone. A shape change of landmark configurations can be most efficiently summarized to the extent that the comparisons can be assumed uniform over extended regions, specifically whole triangles of three landmarks. Such an approximation will not exactly match the biological homology map whenever the "true" deformation takes interior straight edges to curves. Yet it is a very useful first approximation, guiding the selection of a set of relevant contrasts. For this case the biorthogonal crosses of any comparison of two triangles can be constructed by straightforward Euclidean geometry.

This construction requires a ruler, a right angle, a sharp pencil, and a sheet of paper quite a bit larger than the forms. The construction tends to take less space when the angles of the triangles are not too small or too large. You may also wish to have a compass for drawing a circle, colored pencils for differentiating the lines, and tracing paper.

Select two triangles with homologous landmarks at the vertices of each: six points A, B, C, A', B', C'. We will construct the biorthogonal directions for the deformation carrying triangle ABC to the triangle $A'B'C'$, Figure A.2.1.

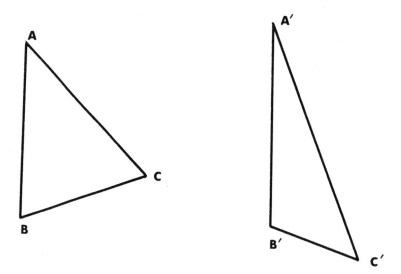

Figure A.2.1 Biorthogonal analysis of triangles by hand, Step 1.

A.2 Biorthogonal Analysis of Triangles by Hand

Step 1, Figure A.2.2. Choose any two landmarks on the first triangle, here A and B. Draw the second triangle ($A'B'C'$) over the first so that landmarks A and A' are superimposed (i.e., set $A = A'$) and so that landmark B' is in line with edge AB (extended if necessary).

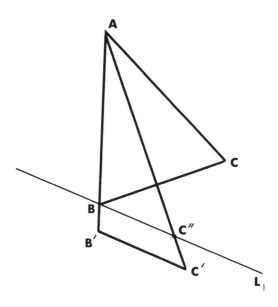

Figure A.2.2 Steps 1 and 2.

Step 2, Figure A.2.2. Draw the line L_1 parallel to $B'C'$ through B, and construct point C'', its point of intersection with edge AC' of triangle $A'B'C'$ (extended if necessary).

Step 3, Figure A.2.3. Draw line L_2 through points C and C'' and locate point F, the midpoint of the line segment between points C and C''.

Step 4, Figure A.2.4. Construct the line L_3 perpendicular to L_2 through F. Each point of line L_3, the **perpendicular bisector** of the segment CC'', is equidistant from points C and C''.

Step 5, Figure A.2.5. Construct the point Q at which the line L_3 intersects the line AB (extended if necessary). If Q is a long way from A and B (i.e., if line L_3 is nearly parallel to line AB) so that X_1 of X_2 (see Step 6) is off the paper, there is no need to locate Q precisely (use alternate steps 6a and 7a).

Step 6, Figure A.2.6. Draw the circle H with Q as center and passing through C and C''. H intersects the line AB at two points X_1, X_2.

Step 6a, Figure A.2.6 (right). If point Q is inaccessible, construct a line, L_4, perpendicular to AB through point F. The intersection of lines L_4 and AB is point W.

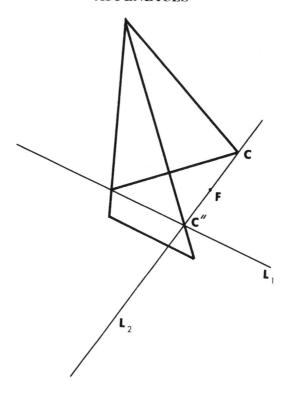

Figure A.2.3 Step 3.

Construct point Y at the intersection of lines AB and L_2. Construct point X halfway between points W and Y on line AB. (For C and C'' close together, this is an approximate location of the correct X that would be constructed upon sufficiently large paper.)

Step 7, Figure A.2.7. The biorthogonal lines we seek for triangle ABC are along the directions of the lines CX_1, CX_2. The biorthogonal lines for $A'B'C'$ are along the directions of $C''X_1$, $C''X_2$. At this time, check that these axes are in fact perpendicular to each other (at least within a range of 89° to 91°). If not, recheck your construction, especially circle H.

Step 7a. The biorthogonal lines for ABC are parallel to line CX and perpendicular to it, and those for $A'B'C'$ are parallel and perpendicular to line $C''X$.

Step 8, Figure A.2.8. Construct for each triangle (ABC, $A'B'C'$ from Step 1) its own set of axes parallel to the directions just discovered. Each axis should connect one vertex of the triangle to a point on the opposite edge. As a check on the construction of these axes, the intersected edges should be divided into the same proportions in ABC and in $A'B'C'$. We now have the biorthogonal crosses in their triangles. We still need the dilatations, the ratios of homologous axes from triangle to triangle (Step 9), and a more convenient display of the crosses (Step 10).

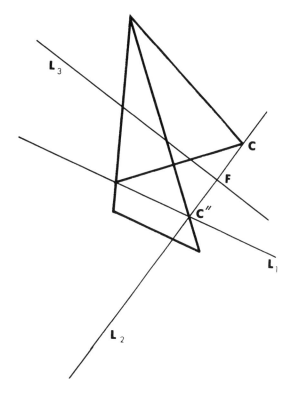

Figure A.2.4 Step 4.

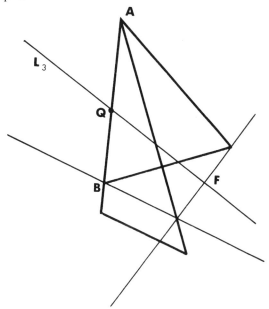

Figure A.2.5 Step 5.

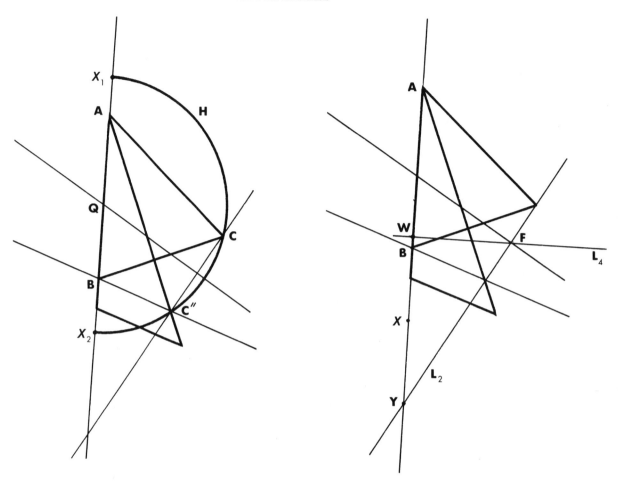

Figure A.2.6 Step 6.

Step 9, Figure A.2.9. Measure the length of one of the biorthogonal axes (from vertex to edge) in triangle ABC and compute its ratio to the homologous length in triangle $A'B'C'$. Likewise form the ratio of the lengths of the other two axes. Be consistent in assignment of numerator and denominator in the two ratios. When the two lengths from $A'B'C'$ are each divided by their homologous lengths in ABC you have computed the dilatations for the transformation from ABC to $A'B'C'$. Their reciprocals describe the deformation in the opposite direction, from $A'B'C'$ to ABC.

Step 10, Figure A.2.10. You may wish to present the results in one of the following displays.

Step 10a, Figure A.2.10a. Choose a position for each cross in the approximate center of its triangle. Copy the lines, with their directions and lengths from Step 8, allowing each axis to bisect the length of the other. Their lengths may be reduced by a suitable factor to make the diagram look better. Display the dilatation values on the first form, in this case, ABC.

A.2 Biorthogonal Analysis of Triangles by Hand

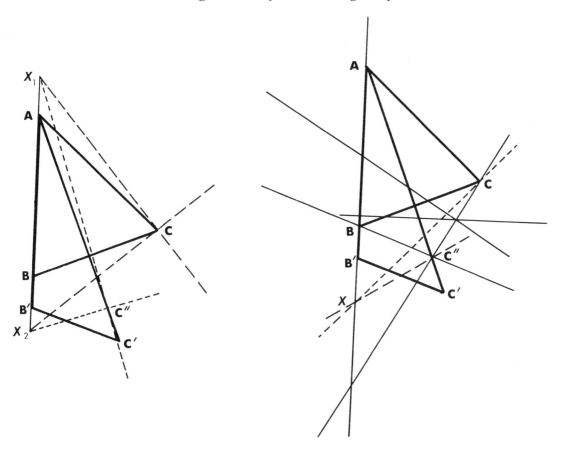

Figure A.2.7 Step 7.

Step 10b, Figure A.2.10b. Draw the two sets of biorthogonal axes from Step 8 on their original triangles. Superimpose the crosses so that each axis will run through a pair of vertices (e.g., A and A', B and B'). The triangles will not be superimposed as they were in Step 1. The dilatations are ratios of distances $A'Z:AZ$ and $B'Z:BZ$, where Z is the cross of the axes.

Proof of the construction. Because the linear transformation $ABC \to ABC''$ leaves points A and B unmoved, it leaves all points of the line AB unmoved. Because X_1 and X_2 are on line AB, they are also left unchanged by the transformation. Because triangles X_1CX_2 and $X_1C''X_2$ are inscribed in a semicircle, they are both right triangles. Thus, the angle between the lines CX_1 and CX_2 remains at 90° during the transformation. Hence these lines must be the biorthogonal axes of the starting triangle ABC, and the lines $C''X_1$, $C''X_2$ into which they are transformed must be the homologous axes after transformation to triangle ABC''. Because triangles ABC'' and $A'B'C'$ are similar, the axes for the transformation of ABC to $A'B'C'$ must be the same as those for the transformation of ABC to ABC''.

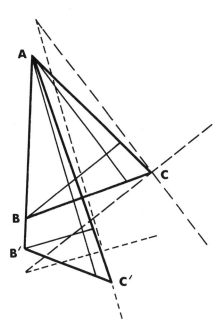

Figure A.2.8 Step 8.

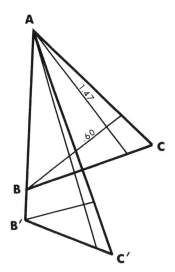

Figure A.2.9 Step 9.

A.3 The Variety of Coordinate Systems   225

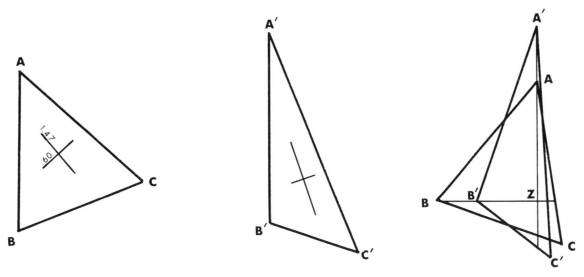

Figure A.2.10 Step 10.

A.3 The Variety of Coordinate Systems

Cartesian coordinates, we noted in Section 3.1.3, are one familiar system for the collection of morphometric archives: perpendicular distances from landmarks to two straight lines at 90°. The Cartesian system is the most familiar of several geometrically simple coordinate systems based on variously specified distances from landmarks to certain special lines or points. These yield different spatial patterns of **coordinate mesh** which may appear more suited to particular biometric investigations.

A.3.1 Other coordinates based in distances to lines

Beyond the perpendicular x- and y-axes of the Cartesian system, there are two other lines that occasionally appear as coordinate curves. In "polar coordinates" (these are misnamed; see below) there are straight lines through the origin. These are lines of constant ratio $y{:}x$, which is to say, constant ratio of distance to one line with respect to distance to another; once again the biological import of such a locus is obscure. Some of the more irritating paradoxes of the allometry literature are based in the fact that passage along regression lines *not* through the origin involves monotone increase or decrease of this coordinate (Fig. A.3.1).

We may also take as our coordinates the set of distances to three lines. If the lines form an equilateral triangle, then the sum of the distances to them is constant throughout the plane. This is a convenient way of plotting three variables that sum to 1, for instance three partial lengths a, b, c as fractions of their sum.

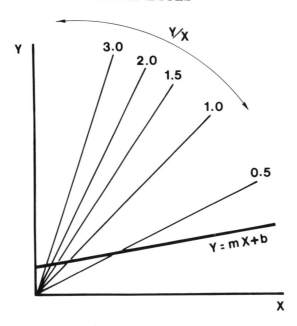

Figure A.3.1 Size allometry of a conventional ratio. The ratio $y:x$ varies monotonically along any allometric regression line not through the origin.

A.3.2 Polar coordinates

By these we mean coordinate systems based on distances to *points* rather than lines. Such systems, though a good deal less familiar even to the mathematician, seem to us much more natural biologically (Bookstein, 1981b). There are several versions.

If we take as our level curves the loci of points whose sum or difference of distances to two fixed points is constant, we get the family of confocal conics (Fig. A.3.2), the set of all ellipses and hyperbolas with the same foci. These will turn out to be the biorthogonal grids for a convenient mapping between quadrilaterals, the projection.

If we set as our level curves the loci of constant sum or difference of distances to *three* points, we get the bicircular quartic curves (Fig. A.3.3), which were the subject of a doctoral thesis in 1890 and have been mostly forgotten since.

A special case of these is the parabolic coordinate system (Fig. A.3.4) in which all the curves are loci for which the distance to a fixed point differs by a constant from the distance to a fixed line through that point.

A.3.3 Mixed systems

If instead of the ratio $y:x$ of Cartesian coordinates—a ratio of distances to *lines*—we instead consider the ratio of the distance to two *points*, the level curves that result

A.3 The Variety of Coordinate Systems

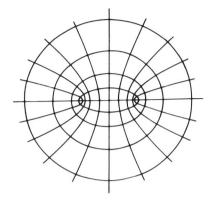

Figure A.3.2 Confocal conics. Sums and differences of distances as coordinates.

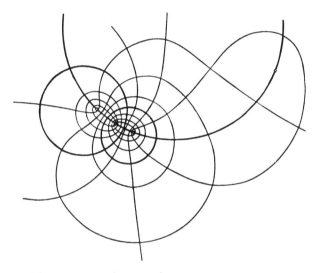

Figure A.3.3 A bicircular quartic coordinate system.

are particular circles around one or the other of those points (Fig. A.3.5), together with the line that is their perpendicular bisector (representing a distance-ratio of exactly 1). It would be nice to continue our tradition of completing the coordinate system with curves at 90° to these. For these circles the orthogonal family is likewise a collection of circles, all the circles through the two points from which we were measuring the distances of the ratio. Circles of this second set are loci of points from which the pair of privileged points spans a constant angle.

These coordinates have two special cases. One, in which the two points fuse into a single center, gives us the parabolic circle coordinates of Figure A.3.6. The other consists in taking one of the points at a very great distance from the region of interest. The circles through the remaining central point become indistinguishable from straight lines. At last we arrive at the ordinary "polar coordinates."

228 APPENDICES

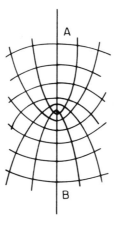

Figure A.3.4 A parabolic coordinate system.

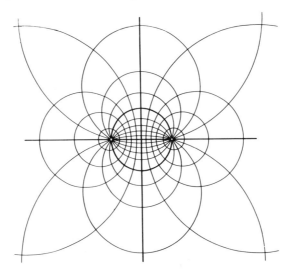

Figure A.3.5 Orthogonal pencils of circles. Angles and ratios of distances as coordinates.

None of these coordinate systems is any more or less unusual or mathematically deceptive than our Cartesian familiars. All come in well-behaved pairs of families intersecting at 90°; all may be generalized to three dimensions. In addition there are entirely different kinds of coordinate systems *for objects which are not points*. Like points, lines, which are of considerable interest to us, may be identified with two coordinates, in a great variety of ways. For instance, we may use the so-called "polar" form of distance from one fixed point and angle from one fixed line (Fig. A.3.7). In most cases these coordinates convert an arc of tangent lines into an arc of points in the appropriate coordinate space, so that we can search for double tangents directly as loci

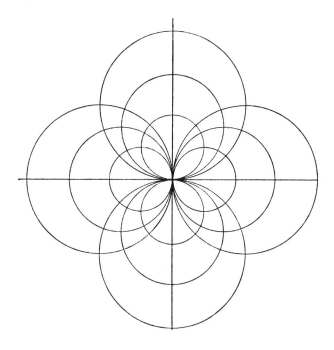

Figure A.3.6 A parabolic circular coordinate system.

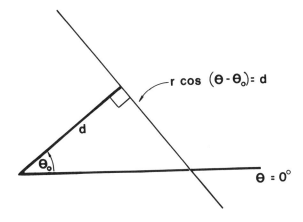

Figure A.3.7 The "polar form," a coordinate system for lines. Set polar coordinates r and θ with reference to an origin O. Any line L may be described by an equation $d = r \cos(\theta - \theta_0)$, where θ_0 is the angle between the line $\theta = 0$ and the normal from O to L and d is the distance from O to L along that normal. The coordinates of L are d and θ_0.

where this curve crosses itself. The further discussion of coordinate systems for lines, and systems of coordinates for more complicated objects such as circles, would take us beyond the scope of this treatise.

A.4 Multivariate Analyses for a Triangle of Landmarks

In emphasizing the effective description of differences or trends in form, this monograph has not dealt much with tests for statistical significance of the differences or trends that are uncovered. This appendix demonstrates a simple but rigorous statistical analysis for these quantifications of form-change: a useful general significance test for shape and growth differences between populations. Designed for triangles of landmarks, the analysis assumes uniform (spatially constant) deformations, the model discussed in Section 2.1. The computations use only Cartesian coordinates of the landmark points rather than requiring that any size or shape variables be specified in advance. Computations can be managed by any of the popular statistical packages.

Statistical analysis of a deformation begins by superimposing triangles on a common baseline, as in Appendix A.2 (Fig. A.4.1). Consider the points A and B of the construction in that appendix to define a Cartesian coordinate system: put point A at the origin $(0,0)$, and set point B at unit distance from A along the x-axis, at the point $(1,0)$. This is surely not the coordinate system in which the triangle $\triangle ABC$ was

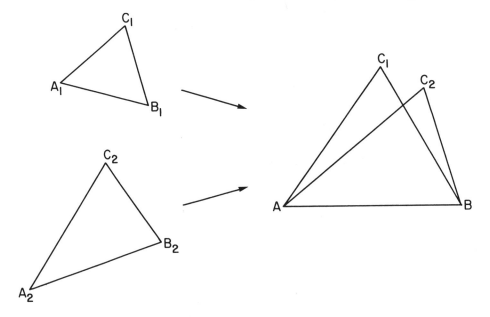

Figure A.4.1 The superposition of two triangles on a common baseline.

A.4 Multivariate Analyses for a Triangle of Landmarks

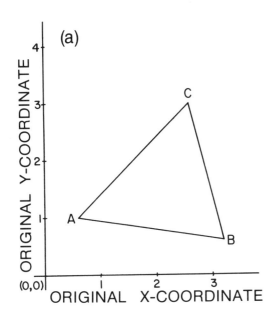
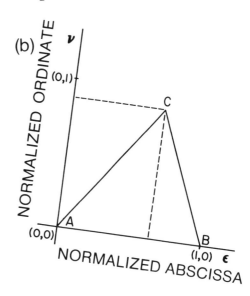

Figure A.4.2 Shape coordinates. (a) The original coordinate system. (b) The shape coordinates (ϵ,ν) of a triangle $\triangle ABC$ are the x- and y-coordinates of one of its landmarks with respect to a Cartesian system translated, rotated, and scaled so that the other two landmarks are at $(0,0)$ and $(1,0)$.

originally digitized (Fig. A.4.2a). That system (perhaps the system of wires of the digitizing tablet) placed landmark A at (x_A, y_A), let us say, B at (x_B, y_B), and C at (x_C, y_C). Switching to the new coordinate system with A at $(0,0)$ and B at $(1,0)$ is the same as working the ruler-and-compass construction of Appendix A.2 with reference to a starting triangle having A at $(0,0)$ and B at $(1,0)$. Using the formulas of Appendix A.1, one can show that the point C which had been digitized at (x_C, y_C) is assigned (see Fig. A.4.2b) the new coordinates (ϵ, ν) where

$$\epsilon = \frac{(x_B - x_A)(x_C - x_A) + (y_B - y_A)(y_C - y_A)}{(x_B - x_A)^2 + (y_B - y_A)^2}$$

$$\nu = \frac{(x_B - x_A)(y_C - y_A) - (y_B - y_A)(x_C - x_A)}{(x_B - x_A)^2 + (y_B - y_A)^2}$$

These formulas, applied to the point A, send it to $(0,0)$ and B to $(1,0)$. In this algebraic form the "construction" may be worked by *MIDAS*, *SPSS*, *SAS*, or any other conventional statistical computing package. We will call ϵ and ν the **shape coordinates** of the triangle $\triangle ABC$. (In terms of complex variables, we have $(\epsilon, \nu) = (C - A)/(B - $

A); Bookstein, 1985.) Even though the coordinates appear to be describing a single landmark, in fact they represent the shape of the triangle as a whole—their statistics are independent of which point was chosen to be C. This will become clearer below.

Consider, now, a *sample* of homologously labelled triangular shapes $\Delta A_i B_i C_i$, where i indexes the separate forms of the sample. If we standardize size and position so that the points A_i are all at (0,0) and the B_i all at (1,0), we have discarded all information about the length of the baseline $A_i B_i$ of the several triangles, so that all possibility of measuring size is lost. But because we have not changed the shapes of the triangles, we have not lost any shape information. In fact we have reduced the shape of each triangle $\Delta A_i B_i C_i$ to the location of a single point, C_i, in the construction registered at $A_i = (0,0)$ and $B_i = (1,0)$.

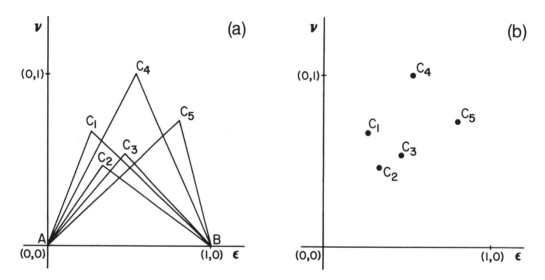

Figure A.4.3 A scatter in the plane of shape coordinates. Once size information is discarded, a sample of triangular shapes $\Delta A_i B_i C_i$ is represented by the scatter of the points C_i when all the A's are sent to (0,0) and all the B's to (1,0) in a single Cartesian system. (a) Triangles after superposition on AB. (b) Equivalent scatter.

In this way the sample of triangular shapes $\Delta A_i B_i C_i$ becomes a scatter of points C_i in the (ϵ,ν)-plane (Fig. A.4.3). If the triangles came in matched pairs $\Delta A_{i1} B_{i1} C_{i1}$, $\Delta A_{i2} B_{i2} C_{i2}$, we would represent the sample of triangular shape changes by a scatter of vectors, connected point-pairs C_{i1}–C_{i2}, as in Figure A.4.4. If the triangular forms came in a series, as in a growth study, there would result extended trajectories, chains of points C_{i1}–C_{i2}–...–C_{ik} following the shape of the triangle ΔABC over the series.

We call ϵ and ν shape coordinates because each itself is a shape variable, that is, a ratio of distances. Triangle by triangle, the vertical coordinate ν is the distance (in either coordinate system) from point C_i to the baseline $A_i B_i$, divided by the length of that baseline, which in the new coordinate system has a length of 1. The horizontal

A.4 Multivariate Analyses for a Triangle of Landmarks

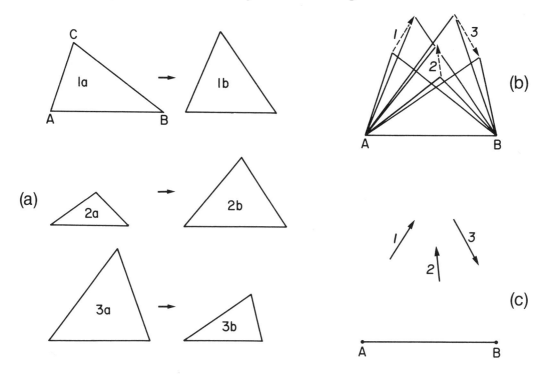

Figure A.4.4 The scatter of pairs of triangles. The change from one form to a matched form is represented as a pair of points in the shape coordinate plane. A set of such changes becomes a scatter of vectors from each form to its match in this plane. (a) The raw data, before scaling. (b) All the triangles superimposed on a common baseline, with matched pairs connected. (c) Suppressing the redundant lines yields a scatter of vectors, one for each shape change.

coordinate ϵ is the proportion in which the foot of the perpendicular from C_i divides the baseline A_iB_i. Like any other set of Cartesian coordinates, these shape coordinates are not themselves homologous measures—they depend explicitly upon perpendicularity to the baseline chosen for the construction.

When one changes baseline, the formulas for the coordinates ϵ and ν appear to change. If we switch from baseline A_iB_i to baseline B_iC_i, for instance, the shape coordinates become those of point A_i after a registration with B_i fixed at $(0,0)$ and C_i at $(1,0)$ by the construction. The scatter of the coordinates (ϵ,ν) that result is shown in Figure A.4.5 for the same triangles that were used in Figure A.4.3.

The effect of changing the choice of baseline is principally to *rotate, translate,* and *rescale* the (ϵ,ν) scatterplot without otherwise altering its form. This result, which is quite unexpected, can be shown to follow from the fact that the construction of shape-change tensors (Appendix A.2) is independent of that same choice of baseline; the deduction is explained in Bookstein (1984a, 1984b). The less the triangles vary in shape, the less the scatter is distorted when the baseline is changed. The invariance of

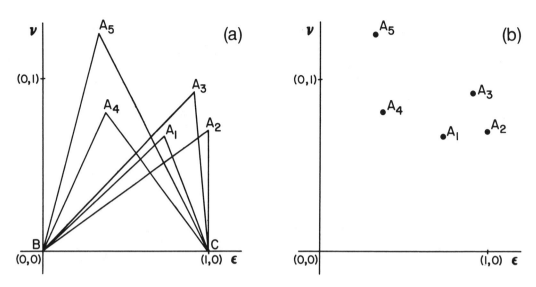

Figure A.4.5 Invariance of the shape coordinate scatter. When the baseline for a shape coordinate scatter is changed, the principal effect upon this diagram is rotation, translation, and rescaling. Multivariate statistical comparisons are not much affected. The scatter here is for the same data as that in Fig. A.4.3, but plotting the position of landmark A upon a baseline from B to C. (a) Triangles after superposition on BC. (b) Equivalent scatter.

the scatter is an approximation more valid as the variance of the shape coordinates becomes smaller. It can be shown also that, when shape variation is moderate, any shape variable measuring the triangle $\triangle ABC$ is statistically equivalent to one of these coordinates ϵ or ν for a baseline using one vertex together with a point dividing the opposite side in a suitable fraction.

Because the scatter of any one set of shape coordinates includes all the shape information available, and is merely moved, rescaled, and rotated by change of baseline, any statistical method for analyzing these scatters which is itself independent of translation, rotation, and rescaling will be nearly independent of the choice of shape and size variables (Bookstein, 1984b). As it happens, although this feature is of little use in most other applications, the general multivariate† linear model and its special cases show exactly the invariance we need under rotation of the vectors of observations. (Their invariance under translation or rescaling is more familiar.) In other words, multivariate analyses of (ϵ,ν) scatters deriving from different baselines result in very nearly the same findings.

For instance, the presence of a mean difference in shape between two groups of triangles (Fig. A.4.6a,b) may be tested in an invariant manner by Hotelling's T^2

---

† In all these applications, the "multivariate" aspect is actually bivariate: the "observations" are ϵ and ν, totaling two variables.

A.4 Multivariate Analyses for a Triangle of Landmarks 235

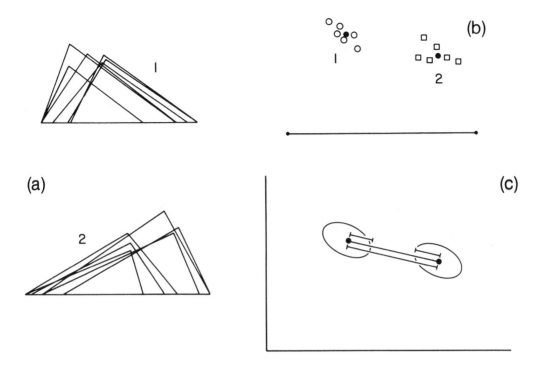

Figure A.4.6 Test of a difference in mean shape. (a) Two populations of triangles. (b) The two scatters of shape coordinates that correspond, each with its centroid. (c) Hotelling's T^2 is a ratio of the distance between two centroids to its within-group standard error in that direction.

statistic. This test, explained in most texts of applied multivariate analysis, compares the length of the vector between the two population centroids in the figure to a denominator expressing the within-group variance in that direction (Fig. A.4.6c). The T^2 statistic is significant at the 5% level when the corresponding Student's t-ratio for the particular direction of best separation is significant at about the 1% level. (This corresponds to a correction by the factor $1/E[\chi_2] = \sqrt{2/\pi}$, about 0.80.) There is also a matched T^2 test for comparing a single population mean with zero. In the present context, this becomes a test for the significance of a shape change observed in a single population—that is, the average of a collection of vectors like that in Figure A.4.4—against the null hypothesis of no mean shape change. The denominator for this T^2-ratio uses the variance of the matched change scores representing the individual shape-change tensors rather than the (generally larger) population variance of the starting or ending forms themselves. In growth studies, individual differences in the times between successive observations of the forms can be adjusted by scaling the lengths of the vectors in the (ϵ, ν) plane (Bookstein, 1984a). Two populations of these mean changes may be compared by a T^2 test of their own, which determines whether the difference between the two samples of vectors of change is larger than that expected from the within-group variance in those same vectors of change. These three applications of Hotelling's T^2 test, all independent of the choice of a baseline for the (ϵ,ν) scatter, are exemplified in Bookstein (1984b).

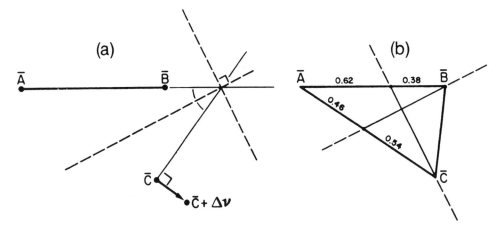

Figure A.4.7 Reporting a mean shape difference. The best report of a mean difference in the plane of shape coordinates uses a ratio of distances at 90° given by an approximation to the construction of Appendix A.2. (a) Construction of the principal directions as bisectors of the angle shown. (b) The principal axes interpreted as a pair of distances measured from one vertex through a point dividing the opposite side in the proportions indicated. One of these distance variables shows the algebraically largest difference of homologous lengths between the mean forms, the other the algebraically smallest difference. The shape variable we seek is the ratio of these two measured lengths.

Whether between groups, between times of observation of a single group, or between changes observed over time in different groups, a mean difference that is significant according to the T^2 test may be optimally expressed using a specific shape variable by completing the construction of Appendix A.2, as indicated in Figure A.4.7. This construction results in a cross of principal axes quite independent of the baseline chosen. In this way we complete our analytic cycle in the same spirit of indifference to a priori variables with which we began it, and report the shape contrast in question without referring to any predetermined list of measures of size or shape separately.

When the Hotelling's T^2 statistic does *not* exceed the appropriate significance level, the two populations being compared cannot be considered to differ significantly in mean shape (or in mean shape-change, whichever was being tested). In this circumstance, ratios of homologously measured lengths, or changes in these ratios, must be considered to be independent of direction. In other words, the size difference between the populations, whether significant or not, must be taken to be isometric, the same in every direction between every pair of landmarks.

Analyses of more than three landmarks result in multiple triangles. Bookstein (1984b) demonstrates the role of the extended biorthogonal grid (Secs. 4.5, 4.6) in

A.4 Multivariate Analyses for a Triangle of Landmarks

drawing the researcher's attention to particular triangles for analysis by this method. The separately optimal shape discriminators may be combined by the path-analytic multiple discrimination model of Section 4.2.

The more general multivariate linear model, of which Hotelling's T^2 is a special case, may exploit the shape coordinates (ϵ,ν) to provide tests of more subtle or complex hypotheses and descriptors. Consider, for instance, size allometry. The covariation of shape with size is displayed upon the (ϵ,ν) plot as a set of contour lines (Fig. A.4.8). These locate shapes having equal predicted values of size in a multiple regression upon the shape coordinates ϵ and ν together. Size allometry is tested by the F-statistic for this multiple regression. It is expressed in a significant regression of size upon the shape coordinates.

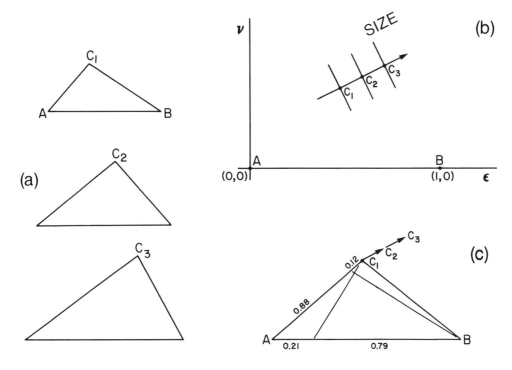

Figure A.4.8 Size allometry in the shape coordinate plane. For a triangle of landmarks, size allometry is unambiguously expressed in the multiple regression of one particular size variable, the sum of the squares of the edge-lengths, upon the shape coordinates jointly. The presence of allometry is tested by the F-ratio for that multiple regression; its geometry is described by the prediction gradient construed as a shape variable. (a) A population of triangles manifesting size allometry. (b) Contours of equal predicted size. (c) The shape variable bearing this allometry, derived as in the preceding figure.

For this test to make sense, it must use a size variable that is uncorrelated with shape whenever there is in fact no size allometry. The discussion of Section 2.2 is relevant here: for most choices of a size variable, the covariances of the shape coordinates with that size measure are non-zero under suitable "null" hypotheses. One useful null model (Bookstein, 1984a), appropriate when the (ϵ,ν) scatter is nearly circular, represents the raw data as derived from a fixed mean triangle ΔABC by the application of identical random noise in each coordinate at each centroid separately. For every mean shape of the triangle of interest, on this null model (which has no size allometry) there is a single size measure uncorrelated with all shape variables. For equilateral triangles, that size variable is ordinary area. For triangles of a mean shape that is not equilateral, the size variable is chosen as the sum of the squares of the sides of the triangle, or, equivalently, as the sum of the sides of the triangle, each weighted by its own population mean length. In research involving multiple populations, multivariate analysis of covariance may be used to test apparent differences in the detailed dependence of size upon the shape coordinates ϵ,ν group by group.

In this appendix we can only hint at the power of these invariant statistical procedures for studies of shape difference, size difference, and allometry. The detailed explanation of hypothesis tests corresponding to the morphometric descriptors suggested throughout this monograph will require a volume of its own.

A.5 Notes on Computations

The analyses in this monograph could not have been produced without the facilities of the University of Michigan computing center and its operating system MTS. We made use of specific programs as follows.

Our digitizing, at a 15" by 15" SummaGraphics digitizing tablet accurate to .002", was managed by the program *ACQUIRE* written by Mr. R. L. Miller of the Center for Human Growth and Development at the University of Michigan.

All the computed line drawings were produced on a CalComp plotter under Fortran program control via calls to the graphics application package *IG* (*Integrated Graphics*), written by Andrew Goodrich of the University of Michigan Computing Center.

Our conventional statistical analyses—regression, principal components, discriminant analysis—were executed in 64-bit arithmetic (double precision) by the program package *MIDAS* written by Daniel J. Fox of the Statistical Research Laboratory, University of Michigan.

The factor analyses of Section 4.2.3 were executed by an unpublished program written by F. L. B. and R. E. S. after the algorithm published in Wright (1954, 1968). Its core loop, for computation of the primary size factor, is presented in Appendix A.5.1.3.

The fitting of distance-trusses by quadrilaterals in a plane (Sec. 3.2.2) is carried out by another unpublished program written by F. L. B. and R. E. S. It makes use of a

A.5 Notes on Computations

proprietary nonlinear optimization subprogram *ZXSSQ* from the IMSL program library.

The medial axes of Section 4.1 were computed by an experimental program, *LINESKEL*, set up explicitly for polygonal data. Its algorithm was published in Bookstein (1979).

The biorthogonal grid computations were all managed by F. L. B.'s large program package *BIORTH* running under MTS. The package has four parts: input of curving outline forms, interpolation from outline homologies and derivation of biorthogonal crosses, integration of crosses into grids, and output of various statistics and images. The second and third of these functions were explained in Bookstein (1978a, 1978b). Graphical output is delegated to the *IG* subsystem mentioned above. Curving forms are specified to the system as landmarks connected by lines or conic arcs. These, in turn, are described in terms of point-coordinates acquired by prior digitization or by explicit manipulation of the reconstructed image at the computer terminal. Their combination into closed curves is managed by the *BIORTH* control language, whose capabilities were sketched in Section 4.5. More detailed documentation of the features of this program is available from F. L. B.

Another of Dan Fox's programs, *TEXTEDIT*, handled the word processing necessary for assembling this book out of its fragments.

Support for all this was derived, at one time or another, from each of the following sources:

NSF Grant DEB-8011562 to R. E. S. and G. R. S.;
NIDR grant DE-03610 to R. E. Moyers;
NIDR grant DE-05410 to F. L. B.;
NSF Grant BSR-8307719 to R. E. S. and F. L. B.

A.5.1 Three program listings

This section lists three algorithms useful to readers wishing to try some of the techniques we recommend. Our program for medial-axis analysis extends over 1000 lines of Fortran, that for biorthogonal analysis over 5500; clearly they could not be included.

A.5.1.1 Principal directions and dilatations.—The following Fortran subroutine computes exact principal dilatations and directions corresponding to the geometric construction of Appendix A.2. Note the three distinct types of error return.

APPENDICES

```
      SUBROUTINE STRAIN(Z1, Z2, ZPRIN, PS, TOL, IER)
C
C     Fred L. Bookstein
C     Center for Human Growth and Development
C     University of Michigan
C     May, 1984
C
C     This subroutine computes the principal strains for the
C        transformation of one triangular element into another.
C     It follows the algorithm published in Figure 6 of
C        F. L. Bookstein, "A Statistical Method for Biological
C        Shape Comparisons," Journal of Theoretical Biology,
C        v. 107, pp. 475-520, 1984.
C
      IMPLICIT COMPLEX*8(Z)
      DIMENSION Z1(3), Z2(3), ZPRIN(2), PS(2), TOL(2)
C
C     Arguments:
C        Z1 -- three complex numbers representing the vertices
C           (x- and y-coordinates) of a first triangle
C        Z2 -- three more complex numbers for the vertices of
C           a second triangle
C        TOL -- a vector of two small positive numbers representing
C           tolerances for various error conditions (see IER)
C
C     Outputs:
C        ZPRIN -- a pair of complex numbers representing directions
C           of principal strain.  Each ZPRIN is of unit length.
C           The vector ZPRIN(1) is the direction of larger principal
C           strain in the plane of the vertices Z1(1...3); the
C           strain along this direction is PS(1).  The vector
C           (0,1)*ZPRIN(1), perpendicular to ZPRIN(1), bears the
C           smaller principal strain, whose magnitude is PS(2).
C           The vector ZPRIN(2) is the direction corresponding to
C           ZPRIN(1) in the plane of the second triangle Z2(1...3).
C           If the shape change is isotropic (PS(1)=PS(2)), then both ZPRIN
C           are set to (0.,0.).
C        PS -- a pair of real numbers embodying the two
C           principal strains.
C            If PS(1)>PS(2), then PS(1) is the principal strain along
C             ZPRIN(1) and PS(2) is the (lesser) principal strain
C           along ZPRIN(1)*(0,1).
C           If PS(1)=PS(2), then the transformation is isotropic and both
C           entries in ZPRIN are (0,0).
C        IER -- a return code.
C           0 -- normal return, ZPRIN and PS loaded appropriately
```

A.5 Notes on Computations

```
C           1 -- transformation is isotropic (ratio of greater to lesser
C              principal strain is less than 1+TOL(1)); returns equal
C              entries in PS and null ZPRIN
C           2 -- no computation, one triangle or the other has ratio of
C              height to base less than TOL(2)
C           3 -- no computation, the triangles have areas of opposite sign
C
C     Three handy functions:
      ZTHREE(ZA,ZB,ZC) = (ZC - ZA) / (ZB - ZA)
C        position of ZC after ZAis moved to (0,0) and ZB to (1,0)
      QNORM(Z) = SQRT(REAL(Z*CONJG(Z)))
C        length of Z
      ZNORM(Z) = Z / CMPLX(QNORM(Z),0.)
C        unit vector along Z
C
      DIMENSION BLENTH(2), ZAFF(2)
      BLENTH(1) = QNORM(Z1(2) - Z1(1))
      BLENTH(2) = QNORM(Z2(2) - Z2(1))
C
C     lengths of the baselines, vertex 1 to vertex 2,
C        in the two triangles
C
      ZAFF(1) = ZTHREE(Z1(1),Z1(2),Z1(3))
      ZAFF(2) = ZTHREE(Z2(1),Z2(2),Z2(3))
C
C     these are the points C, C'' of Figure 6 in Bookstein, op. cit.
C
      IF (AIMAG(ZAFF(1))*AIMAG(ZAFF(2)) .LE. 0.) GO TO 50
      IF (ABS(AIMAG(ZAFF(1))) .LT. TOL(2) .OR. ABS(AIMAG(ZAFF(2))) .LT.
     &    TOL(2)) GO TO 40
      ZDISP = ZAFF(2) - ZAFF(1)
      ZAFFM = (ZAFF(1) + ZAFF(2)) / (2.,0.)
      QDISP = QNORM(ZDISP)
      IF (QDISP/ABS(AIMAG(ZAFFM)) .LT. TOL(1)) GO TO 30
      IF (ABS(REAL(ZDISP)) .GT. .01*ABS(AIMAG(ZDISP))) GO TO 10
      XINT = REAL(ZAFF(1)) - (REAL(ZDISP)/AIMAG(ZDISP)) * (AIMAG(ZAFF(1)
     &   )**2/AIMAG(ZAFF(1) + ZAFF(2)))
C
C     second-order approximation for very large circles
C        (follows from Law of Cosines)
C
      GO TO 20
   10 XCENT = REAL(ZAFFM) + AIMAG(ZDISP) * AIMAG(ZAFFM) / REAL(ZDISP)
C
C     x-coordinate of the intersection between the real axis and
C        the perpendicular to the segment from ZAFF(1) to ZAFF(2)
```

```
C         at its midpoint;
C      center of the circle in Figure 6c of Bookstein, op. cit.
C
       XINT = XCENT + QNORM(CMPLX(XCENT,0.) - ZAFF(1))
C
C      one intersection with the real axis of the circle
C         perpendicular to that axis and passing through
C         ZAFF(1) and ZAFF(2)
C
   20 ZPRIN(1) = ZAFF(1) - CMPLX(XINT,0.)
      ZPRIN(2) = ZAFF(2) - CMPLX(XINT,0.)
C
C      directions of one principal strain relative to the baseline
C         in this construction
C
      Q1 = QNORM(ZPRIN(1))
      Q2 = QNORM(ZPRIN(2))
      ZPRIN(1) = ZPRIN(1) * ZNORM(Z1(2) - Z1(1)) / CMPLX(Q1,0.)
      ZPRIN(2) = ZPRIN(2) * ZNORM(Z2(2) - Z2(1)) / CMPLX(Q2,0.)
C
C      rotate each axis back into its own coordinate frame
C
      PS(1) = (Q2/Q1) * (BLENTH(2)/BLENTH(1))
      PS(2) = (AIMAG(ZAFF(2))/AIMAG(ZAFF(1))) / (Q2/Q1)
      PS(2) = PS(2) * (BLENTH(2)/BLENTH(1))
      IER = 0
      IF (PS(1) .GT. PS(2)) RETURN
      X = PS(1)
      PS(1) = PS(2)
      PS(2) = X
      ZPRIN(1) = ZPRIN(1) * (0.,1.)
      ZPRIN(2) = ZPRIN(2) * (0.,1.)
      RETURN
   30 ZPRIN(1) = (0.,0.)
      ZPRIN(2) = (0.,0.)
      PS(1) = BLENTH(2) / BLENTH(1)
      PS(2) = BLENTH(2) / BLENTH(1)
      IER = 1
      RETURN
   40 IER = 2
      RETURN
   50 IER = 3
      RETURN
      END
```

A.5 Notes on Computations

A.5.1.2 A SAS procedure for shearing.—The following *SAS* commands shear a pooled within-group size factor from principal components two and three. We are grateful to Les Marcus of Queens College, New York, for an early draft of this routine; the version reprinted here was coded and tested by David Swofford of the Illinois Natural History Survey, Champaign-Urbana, Illinois.

The input variable list should consist of a grouping variable followed by the morphometric variables untransformed. The cases must be presorted by the grouping variable, which must take on consecutive values beginning with 1. No missing values are permitted.

```
*-----Modify the next four lines as appropriate--------------;

%LET         NVAR      = 10;
%LET         INPUT     = TEST.DATA;
%LET         OUTPUT    = SHEAROUT;
%LET         RETAIN    = 3;

*   where NVAR    = the number of morphometric variables
                    to be included in the analysis,
          INPUT   = the name of the input data set,
          OUTPUT  = the name of the SAS data set to which
                    principal component scores and log
                    transformed variables are to be written
          RETAIN  = the maximum number of principal components
                    to be computed (only the second and third
                    will be sheared in this example);


*--------------Data set for output labelling-----------------;

Data PCLAB;
   LENGTH LAB1-LAB&NVAR $ 8;
   ARRAY LAB(I) $ LAB1-LAB&NVAR;
   DO I = 1 TO &NVAR;
      LAB = {sp}PC_{sp};
      SUBSTR(LAB,3) = I;
      LAB = COMPRESS(LAB);
   END;

PROC MATRIX;
   FETCH X DATA=&INPUT COLNAME=VARLAB;
   VARLAB = VARLAB{sp}(2:&NVAR+1,);
   NIND = NROW(X);                        * NIND = number of
                                                   individuals;
   NGRP = X(NIND,1);                      * NGRP = number of
                                                   groups;
```

APPENDICES

```
    D = DESIGN(X(,1));
    N = D(+,);                                  * N = row vector of
                                                    group sample sizes;
                                                    sizes;

    GROUP = X(,1);
    X = LOG10(X(,2:&NVAR+1));                   * delete 1st column of
                                                    X, transform to logs;
    C = X(.,);                                  * C = grand centroid;
*---------Usual principal components analysis------------;

    Q = (X{sp}*X - C{sp}*C*NIND)#/(NIND-1);     * Q = total covariance
                                                    matrix;
    EIGEN A E Q;                                * A = eigenvalues,
                                                    E eigenvectors of Q;
    RETAIN = MIN(&NVAR||&RETAIN);
    E = E(,1:RETAIN);
* COMPUTE EIGENVALUE PROPORTIONS;
    A = A||A||A;              --                * make room;
    SUMVAL = A(+,1);
    ASUM = 0;
    DO I = 1 TO &NVAR;
       A(I,2) = A(I,1)#/SUMVAL;
       ASUM = ASUM+A(I,2);
       A(I,3) = ASUM;
    END;

    FETCH PCLAB DATA=PCLAB TYPE=CHAR;           * get labels created
                                                    above;
    PC = X * E;                                 * PC = pc scores;

*--------Compute within-group covariance matrix (Q{sp})---------;

    IHOLD = 0;
    DO I = 1 TO NGRP;
       NI = N(1,I);
       XI = X(IHOLD+1:IHOLD+NI,);                * partition X by group;
       IHOLD = IHOLD + NI;
       CI = XI(.,);                              * centroid of
                                                    Ith group;
       XCI = XI - (J(NI,1) * CI);                * center data for Ith
                                                    group;
       XC = XC//XCI;                             * XC = group-free
                                                    X-matrix;
    END;
```

A.5 Notes on Computations 245

```
QP = XC{sp} *XC #/ (NIND-1);            * QP = pooled within-
                                          group covariance
                                          matrix Q{sp};

*--------Shear 2nd and 3rd principal components------------;

EIGEN AP EP QP;                         * AP = eigenvalues,
                                          EP = vectors of Q{sp};
PCZ = XC * E(,1:3);                     * translate original
                                          PC-scores to 0 mean
                                          within groups to
                                          yield PC(z);
S = -EP(,1);                            * S = pooled within-
                                          group size factor;
PCS = XC * S;                           * PCS =  scores on
                                          size factor S;

DENOM = PCS{sp} * PCS;                  * regress PC(z)
                                          scores on S scores;
ALPHA2 = (PCS{sp} * PCZ(,2))# DENOM;    * (no intercept is
                                          needed since all
ALPHA3 = (PCS{sp}* PCZ(,3))# DENOM;     * means are zero);

PCZ2 = PCZ(,1 2);
PCZ3 = PCZ(,1 3);                       * multiple regression
                                          of S-scores on
BETA2 = INV(PCZ2{sp}*PCZ2)*PCZ2{sp}*PCS;  * PC(z) scores;
BETA3 = INV(PCZ3{sp}*PCZ3)*PCZ3{sp}*PCS;

H2 = E(,2) - ALPHA2*E(,1 2)*BETA2;      * H2 = loadings of
                                                sheared PC 2;
H3 = E(,3) - ALPHA3*E(,1 3)*BETA3;      * H3 = loadings of
                                                sheared PC 3;
HS = X* (H2||H3);                       * HS = scores on
                                          sheared components;
```

At this point there should follow routines to output appropriate tabular listings and plots.

A.5.1.3 Factor analysis using Wright's algorithm.—This Fortran subroutine extracts factors from a correlation matrix according to the model described in Section 4.2.

```fortran
      SUBROUTINE WRIGHT (ICMD1,ICMD2,NVAR,CORR,RCORR,
     &                   FACTOR,IT,LISTVR)
C
C     Subroutine for a Wright-style analysis.  Written by:
C
C       Fred. L. Bookstein
C       Center for Human Growth and Development
C       University of Michigan
C       May, 1981
C
C     Modified by R. E. Strauss (August 1984)
C     Source language:  FORTRAN-77
C
C
C     ---------------------------------------
C
C
C     This subroutine extracts one or two factors (sets of
C     loadings) from a correlation matrix using the method
C     of Wright (1954).
C
C     Arguments received:
C
C       ICMD1 - command indicating action to be performed:
C          0 - partial newly estimated factor from matrix;
C          1 - partial given factor (in 'FACTOR') from matrix;
C          2 - partial two factors from matrix, one provided (in
C                'FACTOR') and one newly estimated.
C
C       ICMD2 - command indicating variables to be used:
C          0 - use all variables in estimation of loadings;
C          1 - omit correlations of variable-pairs given (in
C                'LISTVR') from estimation of loadings;
C          2 - estimate loadings from subset of variables
C                (given in 'LISTVR').
C
C       NVAR - total number of variables, =dimensions of 'CORR'
C          and 'RCORR'.
C       CORR - square correlation matrix having 1's on diagonal.
C       FACTOR - optional factor, a previously estimated set
C          of loadings (see description of ICMD1).
C       LISTVR - optional list (subset) of variables or pairs
C          of variables (see description of ICMD2).  Unused
C          portion of array, after list of variables, must be
C          initialized to zeros.
C       IT - maximum number of iterations used to estimate
C          factor.  Set to zero for default.
C
```

A.5 Notes on Computations

```
C      Arguments returned:
C
C         RCORR - residual correlation matrix after factor(s)
C            have been partialled out.
C         FACTOR - newly estimated set of loadings.
C         IT - number of iterations used to estimated factor.
C
C         ---------------------------------------
C
C *    Maximum array sizes arbitrarily dimensioned as N=100.
C *    'LISTVR' would be dimensioned  N*(N-1)/2  to account
C *    for all possible variable-pairs, but in practice
C *    can be quite small.
C
       DIMENSION CORR(100,100),RCORR(100,100),LISTVR(200),
     &           FACTOR(100),FACTSV(100),IVAR(100),
     &           RFACT1(100),RFACT2(100),FPREV(100)
       LOGICAL SKIP(100,100),FINISH
C
C *    Maximum and actual numbers of iterations
C
       IF (IT.GT.0) THEN
          ITER=IT
          IT=0
       ELSE
          ITER=200
       ENDIF
C
C *    Convergence criterion for loadings.
C
       TOL=0.0001
C
C *    Copy factor loadings.  Initialize
C *    variable-omission matrix and list of variables used.
C *    Count variables passed via 'LISTVR'.
C
       DO 110 I=1,NVAR
         DO 100 J=1,NVAR
           IF (I.NE.J) THEN
              SKIP(I,J)=.FALSE.
           ELSE
              SKIP(I,J)=.TRUE.
           ENDIF
 100     CONTINUE
         FACTSV(I)=FACTOR(I)
         FACTOR(I)=0.0
```

```
      110 CONTINUE
          IF (ICMD1.EQ.1) GO TO 550
C
          IF (ICMD2.EQ.2) GO TO 145
          DO 120 I=1,NVAR
             IVAR(I)=I
      120 CONTINUE
          MVAR=NVAR
          IF (ICMD2.EQ.0) GO TO 160
C
      145 LVAR=0
          DO 150 I=1,NVAR
             IF (LISTVR(I).LE.0) GO TO 160
             IF (ICMD2.EQ.2) IVAR(I)=LISTVR(I)
             LVAR=LVAR+1
      150 CONTINUE
          IF (ICMD2.EQ.2) MVAR=LVAR
C
C *   If requested, omit correlations of given pairs of
C *   variables.
C
      160 IF (ICMD2.NE.1) GO TO 250
          DO 200 I=1,LVAR,2
             SKIP(LISTVR(I),LISTVR(I+1))=.TRUE.
             SKIP(LISTVR(I+1),LISTVR(I))=.TRUE.
      200 CONTINUE
C
C *   Estimate loadings from correlation matrix.
C *   Initial guess is that all loadings are equal
C *   to the absolute root-mean correlation (including
C *   self-correlations).
C
      250 SUMCOR=0.0
          SUMN=0.0
          DO 300 I=1,MVAR
             IVI=IVAR(I)
             DO 270 J=1,I
                IVJ=IVAR(J)
                IF (SKIP(IVI,IVJ)) GO TO 270
                SUMCOR=SUMCOR+CORR(IVI,IVJ)
                SUMN=SUMN+1.0
      270    CONTINUE
      300 CONTINUE
          GUESS=SQRT(ABS(SUMCOR/SUMN))
          WRITE (6,310) GUESS
          WRITE (9,310) GUESS
      310 FORMAT ('  ROOT-MEAN-SQUARE CORRELATION=',F7.3)
          DO 390 I=1,MVAR
```

A.5 Notes on Computations

```
            IVI=IVAR(I)
            FACTOR(IVI)=GUESS
            RFACT1(IVI)=1.0
  390   CONTINUE
        IVM=IVAR(MVAR)
C
        DO 500 IT=1,ITER
          DO 400 I=1,NVAR
            FPREV(I)=FACTOR(I)
  400     CONTINUE
          DO 410 I=1,MVAR
            IVI=IVAR(I)
            RFACT2(IVI)=0.0
            DO 405 J=1,MVAR
              IVJ=IVAR(J)
              IF (.NOT.SKIP(IVI,IVJ)) THEN
                 RFACT2(IVI)=RFACT2(IVI)+(RFACT1(IVJ)*CORR(IVI,IVJ))
                ELSE
                 RFACT2(IVI)=RFACT2(IVI)+(RFACT1(IVJ)*FACTOR(IVI)
     &                        *FACTOR(IVJ))
              ENDIF
  405       CONTINUE
  410     CONTINUE
          R=RFACT2(IVM)
          DO 420 I=1,MVAR
            IVI=IVAR(I)
            RFACT2(IVI)=RFACT2(IVI)/R
  420     CONTINUE
          S1=0.0
          S2=0.0
          DO 430 I=1,MVAR
            IVI=IVAR(I)
            R=RFACT2(IVI)
            IF (.NOT.SKIP(IVI,IVM)) THEN
                S1=S1+(R*CORR(IVI,IVM))
              ELSE
                S1=S1+(R*FACTOR(IVI)*FACTOR(IVM))
              ENDIF
            S2=S2+(R*R)
  430     CONTINUE
          FACTOR(IVM)=SQRT(ABS(S1/S2))
          FINISH=.TRUE.
          DO 450 I=1,MVAR
            IVI=IVAR(I)
            RFACT1(IVI)=RFACT2(IVI)
            IF (IVI.NE.IVM)
     &         FACTOR(IVI)=RFACT2(IVI)*FACTOR(IVM)
            IF (ABS(FACTOR(IVI)-FPREV(IVI)).GE.TOL) FINISH=.FALSE.
```

```
      450   CONTINUE
            IF (FINISH) GO TO 525
      500 CONTINUE
      525 CONTINUE
C
C *   Partial a single factor from the correlation matrix.
C
          IF (ICMD1.EQ.2) GO TO 700
      550 DO 600 I=1,NVAR
            IF (ICMD1.EQ.0) RFACT1(I)=FACTOR(I)
            IF (ICMD1.EQ.1) RFACT1(I)=FACTSV(I)
      600 CONTINUE
          DO 650 I=1,NVAR
            DO 630 J=1,NVAR
              RCORR(I,J)=CORR(I,J)-(RFACT1(I)*RFACT1(J))
              RCORR(J,I)=RCORR(I,J)
      630   CONTINUE
      650 CONTINUE
          GO TO 1000
C
C *   Partial two factors simultaneously from the matrix.
C
      700 S11=0.0
          S12=0.0
          S22=0.0
          DO 740 I=1,NVAR
            DO 730 J=1,NVAR
              C=CORR(I,J)
              S11=S11+(C*FACTOR(I)*FACTOR(J))
              S12=S12+(C*FACTOR(I)*FACTSV(J))
              S22=S22+(C*FACTSV(I)*FACTSV(J))
      730   CONTINUE
      740 CONTINUE
          CORR12=0.0
          IF (S11.GT.0.0 .AND. S22.GT.0.0) CORR12=S12/SQRT(S11*S22)
          DO 800 I=1,NVAR
            DO 780 J=I,NVAR
              RCORR(I,J)=CORR(I,J)-(FACTOR(I)*CORR12*FACTSV(J))
     &                            -(FACTOR(J)*CORR12*FACTSV(I))
              RCORR(J,I)=RCORR(I,J)
      780   CONTINUE
      800 CONTINUE
C
     1000 RETURN
          END
```

Glossary

Affine transformation—a reversible point-to-point transformation leaving straight lines straight and parallel lines parallel. The ratios of distances separating three points on a straight line remain unchanged.

Allometric coherence—the joint application of a **primary size factor**, **secondary factors**, and **unique variance** in a **path analysis** accounting for the covariances of **variables** during growth. This concept is a joint generalization of **growth gradients** and **morphological integration**.

Allometric factor model, allometric model—a **path analysis** in which covariances of the logarithms of measured distances are explained by a single **general size factor**. In this model, the ratio between expected increments in structures of different sizes, divided by the ratio of their sizes, remains constant across size.

Allometric reassociation—changes in the **path coefficients** of **allometric coherence** over evolutionary time; especially, rearrangement of the pattern of **loadings** upon **secondary factors**. Compare **Dissociation**.

Allometry—the dependence of shape variables upon size variables (Gould, 1966).

Analytical boundary—a sequence of **conic arcs** or straight segments connecting **landmarks** and perhaps **pseudolandmarks**, replacing an **empirical boundary** for the representation of a **boundary curve**. The **homology map** is defined on and inside the analytical boundary.

Anatomical homology—assignment of **biological homology** based primarily on specialized properties of tissues or juxtapositions of tissues. Compare **Material homology**.

Archive—a collection of geometric information about a series of biological outline forms, containing coordinates of landmarks and either equations for outline arcs between landmarks or a sample of points from those arcs.

Basis—a set of **variables** whose **linear combinations** are of interest.

Biological homology—a spatial or ontogenetic correspondence among definable structures or parts of two or more different organisms. Assignments of biological homology are supported by various forms of experimental or theoretical argument beyond the scope of this book. See **Anatomical homology**, **Material homology**.

Biorthogonal axes, biorthogonal directions—see **Principal axes**.

Biorthogonal grids—a pair of **coordinate meshes**, one for each form of a **form-comparison**, which correspond according to some **homology map** and are aligned with the **principal axes** of the transformation between the forms. Each grid is at 90° at all of its **mesh points**. The biorthogonal grid is one way of diagramming a **tensor field**.

Bipolar component or **factor**—a **principal component** or **factor** with most loadings nearly equal in magnitude but inconstant in sign.

Boundary curve—the edge of the image of a biological form, modeled as a curve of points in the plane.

Boundary homology function—the extension of the correspondence of **landmarks** to the arcs of an **analytical boundary**; the first step in the computation of a **homology map**.

Boundary landmark—a **landmark** used in representing the **analytical boundary** of a biological form.

Box truss—a network of distance measures among homologous landmarks on a form, patterned as a series of contiguous quadrilaterals having both internal diagonals.

Canonical analysis—one type of discriminant analysis: the optimization, over the **linear combinations** of a **basis**, of variance between groups divided by variance within groups.

Cartesian coordinate pairs—distances to two fixed perpendicular straight lines.

Character—any quantity or code assigned to each of a group of organisms. See **Variable**.

Character space—a geometric representation of the values of **variables** as **Cartesian coordinates** in a Euclidean space of many dimensions. Each point in this space represents one specimen or set of observations. Compare **Dual space**.

Collinearity—a property of a **basis**, that all correlations between pairs of its variables are high. Collinearity engenders uncertainty in the computed coefficients of **linear combinations**.

Communality—that fraction of a variable's variance that is shared with the other latent or observed variables of a **path analysis**. Compare **Unique variance**.

Complexity—sum of **scale-complexity** and **proportion-complexity** over a form or region; two-dimensional variance of a scatter in the ($\log d_1$, $\log d_2$)-plane.

Computed-homologues—points that correspond under a **computed-homology**.

Computed-homology—an extension of the notion of **homology** from **landmarks** to the other points inside or upon a set of outlines. Mathematically, a computed-homology is an interpolation of the correspondence of landmarks according to some algebraic rule.

Confirmatory factor analysis—the augmentation of **factor analysis** by certain assumptions about noise so as to permit statistical tests of hypotheses about **loadings**.

Conic arc—a fragment of a plane curve describable in Cartesian coordinates by an equation of the form $Ax^2 + Bxy + Cy^2 + Dx + Ey + F = 0$. This will represent an arc of a circle, ellipse, parabola, or hyperbola. Conic arcs, along with straight segments, connect **landmarks** into the **analytic boundary** representing a biological form.

Consistency criterion—one of various algebraic equations relating direct effects and observed covariances in a **path analysis**. In **factor analysis**, the equations specify that the coefficients of a **linear combination** of variables be the covariances of the variables with the linear combination. In regression, the equations specify that the sum of direct and indirect effects relating each predictor to the dependent variable be equal to the covariance observed between them.

Coordinate—one of two quantities fixing the position of a point in a plane, or three fixing a point in space.

Coordinate mesh—two sets of curves, each the collection of all points having the same value for one coordinate. When the coordinates are **Cartesian**, for example, the coordinate mesh is ordinary square graph paper.

Cross-sectional data—measures taken once for each **character** on each organism of an analysis. See **Longitudinal data**.

Cursor—a device that indicates the position of a single point upon a **digitizing tablet**.

d_1, d_2—see **Principal dilatations**. The larger dilatation is taken to be d_1.

Deformation—a **model** of **form-change** as **smooth** alteration of position from any point, **landmark** or not, to its **homologue**, observed or computed. See **Tensor field**.

Developmental program—that portion of growth and development that is under genetic, rather than environmental, control.

Digitize—to record coordinates, usually **Cartesian**, of landmarks or other points upon the **geometric form** of an organism.

Digitizing tablet—a device for **digitizing** and simultaneously storing the results in a computing system.

Dilatation—a ratio of lengths of homologous **line-elements** in two forms. Also called, especially in bioengineering, a strain ratio or extension ratio.

Direct effect—in a **path analysis**, the explicit production of change in the value of one variable, observed or latent, by change in another; also, the contribution of such an action to an observed covariance.

Discriminant analysis—any procedure for constructing a **linear combination** of variables that tends to have different values for organisms in different groups.

Discriminant function—the **linear combination** generated by a **canonical analysis**.

Displacement—see **Size displacement**, **Net size displacement**, **Shape displacement**, and **Net shape displacement**.

Dissociation—the evolutionary uncoupling of ancestral associations of **secondary**

factors; a component of **allometric reassociation**. Compare **Regional dissociation**.

Distance—the length of a transect between two landmarks on an organism, used as a **character** or **variable** in morphometric analyses.

Distortion—see **Deformation**.

Dual space—a geometric space in which **linear combinations** of the **variables** of a **basis** may be depicted as vectors out of a common origin. The length of each vector must equal the standard deviation of the variable it represents, and the cosine of the angle between any two vectors must equal the correlation of the corresponding variables. Compare **Character space**.

Eigenvector—a **linear combination** of observed variables which is unaltered, except for scale, under multiplication by some matrix. See **Principal component**.

Empirical boundary—the outline of a form, with its intrinsic irregularities, as it might be drawn or digitized without landmarks.

End point—the terminus of a segment of a **medial axis**, where the two points of tangency to a sequence of circles converge to a single maximum of curvature.

Factor—an unmeasured (**latent**) variable which linearly affects the expected values of a set of observed variables; a common linear cause of the observed variables.

Factor analysis—a class of **path analyses** in which all variables are accounted for as the linear combinations of **factors** together with **unique variance**.

Factor score—a **linear combination** of a set of observed variables. Its value for any individual estimates the value of a **factor** upon which the variables all load.

Form-change, form-comparison—the rigorous description of the difference between two forms; a pair of **geometric forms** together with their relation according to a **biological homology**.

General factor—the single **factor** which best accounts for the set of all observed covariances among variables in a **factor analysis**.

General size—in a **path analysis**, a **general factor** which best accounts for all observed covariances among a set of **distance** measures taken on individuals of varying sizes. See **Primary size factor**.

Geometric form—the formal representation of a **boundary curve** as a sequence of points connected by straight lines or conic arcs. If the points are **landmarks** or **pseudolandmarks**, the geometric form is the **analytic boundary**, and may be augmented by **internal landmarks**. If they are not, it is an **empirical boundary**.

Geometrical deformation—see **Deformation**.

Growth gradient—a consistent increase or decrease, along some curve or axis within a form, in the values of **dilatations**, or of **loadings** on **general** or **primary size**, in the direction of that curve or axis.

Growth trajectory—a function describing some aspect of the size of an organism over chronological or biological time.

GLOSSARY

Helping point—a digitized point which is not a **landmark** but which guides the placement of a **conic arc** during construction of an **analytical boundary**.

Heterochrony—changes in descendent forms in the relative timing of appearance and/or development of ancestral **characters**.

Heteroscedasticity—inhomogeneity or inequality of variances. See **Homoscedasticity**.

Homologues—points in two or more organisms corresponding to one another according to a **homology map**.

Homology—see **Biological homology**.

Homology function, homology map—the extension of the correspondence of **landmarks** to arcs between landmarks and to the interior of the **analytic boundary**: an algorithm which computes, for every point upon or within the analytical boundary of one form, a specific homologous point in another form.

Homoscedasticity—equality of variances of a variable among a set of samples or categories, such as size ranges. Compare **Heteroscedasticity**.

Hyperplane—a Euclidean (non-curving) subspace of dimension one less than the dimension of the space in which it is embedded.

Indirect effect—in a **path analysis**, the action of one variable on another through its correlation with an intermediate variable. Also, the contribution of such an action to an observed covariance.

Internal landmark—a **landmark** positioned inside, rather than upon, the **analytical boundary** of a **geometric form**.

Intrinsic representation—the representation of a geometric form without reference to an underlying coordinate system. For instance, the **box truss** is an intrinsic representation of the configuration of landmarks it measures.

Isometry, isotropy—simple geometric scaling: no change of proportions with change in size.

Kluge-Kerfoot phenomenon—the apparent correlation, across characters, of amount of between-population differentiation with amount of within-population variability.

Landmark—a point (one in each form of a sample) operationally identified by some feature of the local morphology and consistent with some **biological homology**. Compare **Pseudolandmark**.

Latent variable—a **linear combination** of the observed variables, together with unmeasurable noise, that explains certain covariances in a **path analysis**.

Line-element—an arbitrarily small straight-line segment within a biological outline. Its attributes include location, direction, and, in the study of form-change, **dilatation**.

Linear combination—a formula $\Sigma_i a_i X_i$, where the X_i are variables measured over a population and the a_i are coefficients held constant over that population.

Examples: **log size variable**, **log shape variable**, **principal component**, **factor score**, **size-free discriminator**.

Loading—a value proportional to the correlation or covariance of a **variable** with a **factor**. If the loading is positive, the value of the variable increases as the **factor score** increases; if negative, the variable decreases with increasing factor score.

Log shape variable—a **linear combination** $\Sigma_i a_i Y_i$ of the logarithms Y_i of various measured distances for which the sum Σa_i of the coefficients is 0.

Log size variable—a **linear combination** $\Sigma_i a_i Y_i$ of the logarithms Y_i of various measured distances for which the sum Σa_i of the coefficients is 1.

Longitudinal data—measures taken repeatedly on a single individual, or on many individuals, across time. Compare **Cross-sectional data**.

Material homology—assignment of **biological homology** based primarily on identity of substance or ontogenetic continuity of cell lineage. Compare **Anatomical homology**.

Medial axis—the set of centers of circles wholly inside a biological outline and tangent to it at two or more points; the set of points inside a biological outline at the same distance from the outline along two or more distinct normals to the outline.

Meristic variable—a **variable** which is a count of serial or segmental structures. Compare **Distance**.

Mesh point—a point of intersection of curves of a **coordinate mesh**. If the coordinates are Cartesian and equally spaced, the mesh points make up a square grid.

Model—a formalized mathematical expression of postulates and inferences utilized to describe and explain observed data. See **Morphometric model**. Also loosely applied to any reasonably formalized equation, statement, or diagram that is consistent with empirical observations.

Morphological integration—the presence of biologically meaningful **secondary factors** for a sample of organisms.

Morphological space—a geometric **character space** within which the relationships of observed and unobserved forms may be studied.

Morphometric model—the geometrical and statistical computations by which systematic trends or contrasts, in combination with statistical noise (error), account for or reconstruct the original data describing body form and its change.

Morphometrics—the statistical analysis of biological homology treated as a geometrical deformation.

Morphospace—see **Morphological space**.

Multiple discriminant analysis—see **Canonical analysis**.

Net shape displacement—in a scatter of logarithms of **principal dilatations** over a **coordinate mesh**, the perpendicular distance of the centroid from the **size-change axis**; a measure of average **proportion-change**.

GLOSSARY

Net size displacement—in a scatter of logarithms of **principal dilatations** over a **coordinate mesh**, the perpendicular distance of the centroid from the **shape-change axis**; a measure of average **scale-change**.

Normal equations—equations used for computing multiple regressions. They are simultaneous linear equations generated for the unknown coefficients $a_1 \ldots a_n$ of the multiple regression. They are generated by the "least-squares" criterion that the variance of the linear combination $Y - a_1 X_1 - \ldots - a_n X_n$ be minimal, where Y and $X_1 \ldots X_n$ are observed variables. The equations are found by setting derivatives of this variable to zero.

Orthogonal component—any of the terms in an **orthogonal decomposition**, or the coefficient multiplying the special function included in such a term.

Orthogonal decomposition, orthogonal functional analysis—the approximation of an empirical numerical function, such as a **radius function** or a **width function**, by the sum of multiples of special functions which are orthogonal, i.e., such that the integrals of their products are zero over the domain in which the functions are defined.

Outline—see **Boundary curve**.

Paedomorphosis—a type of **heterochrony** in which adult **characters** or stages of a descendent form are homologous with pre-adult characters of an ancestor, presumably due to truncation or deceleration of developmental processes.

Path analysis—**modeling** of a covariance matrix in terms of **latent variables** and their covariances, **loadings** of the observed variables upon the latent variables, **direct effects** of the observed or latent variables upon one another, and **unique variances** of the observed variables.

Path coefficient—a coefficient in a **path analysis**: either a **loading** or a **direct effect** relating latent variables or observed variables.

Path diagram—a diagram of a **path analysis**, in which **direct effects** are indicated by one-headed arrows between variables observed or latent, and two-headed arrows indicate observed covariances or correlations that are not further accounted for.

Peramorphosis—a type of **heterochrony** in which adults of a descendent form have developed characters or stages not present in an ancestor, presumably by acceleration or extrapolation of developmental processes.

Perpendicular bisector—for a segment with endpoints A, B, the set of points equidistant from A and B: the line through the midpoint of A and B perpendicular to the segment connecting them.

Point—a geometrical abstraction representing location without extent. (Euclid.)

Polar coordinates—any of a class of coordinate systems based on distances to fixed points. Compare **Cartesian coordinates**.

Pooled among-group regression—a regression of one or more dependent variables on a predictor variable, treating all observations as if they were drawn from a single population. Compare **Pooled within-group regression** and **Separate within-group regressions**.

Pooled within-group regression—a regression of one or more dependent variables on a predictor variable, performed after subtracting the appropriate group mean from each variable for each observation and then treating all observations as if they were drawn from a single population. Compare **Pooled among-group regression** and **Separate within-group regressions**.

Primary factor, Primary size factor—a modification of **general size** so as to account for only a specified subset of the covariances observed among a set of size measures. Compare **Secondary factor**.

Principal axes—the **line-elements** of largest and smallest **dilatation** with respect to their **homologues**, considering dilatation as a function of direction through a particular pair of points in a **form-comparison**. These directions are perpendicular in both forms.

Principal component—an **eigenvector** of a covariance or correlation matrix.

Principal dilatations—at any point in a form undergoing **deformation**, the **dilatations** along the **principal axes**.

Principal directions—see **Principal axes**.

Proportion-change—the logarithm of the ratio between the **principal dilatations** at a point for a particular comparison of forms, or its average over a form or region. Compare **Scale-change**.

Proportion-complexity—variance of **proportion-change** over a form or region.

Pseudolandmark—a point bearing a reliable operational definition but which is not a **landmark** homologous from form to form. Example: the endpoints of a minimum or maximum diameter of a form.

Radius function—for a closed **geometric form**, the function describing the distance to the outline from an appropriate center point as a function of angle from a suitably chosen starting vector.

Regional dissociation—the developmental independence of spatially defined regions of a biological form. Compare **Dissociation**.

Residual—the difference between an observed value or coefficient and its estimated value according to some **model**.

ρ**F-sets** - sets of variables, highly correlated in a population, whose high correlations can be easily explained, as by functional arguments.

Scale-change—logarithm of the product of the **principal dilatations** at a point for a particular **form-comparison**, or its average over a form or region. Compare **Proportion-change**.

Scale-complexity—variance of **scale-change** over a form or region.

Score—see **Factor score, Linear combination**.

GLOSSARY

Secondary factor—a **factor** computed by a **path analysis** to account for certain covariances left unexplained by a **primary size factor** or other modified **general factor**.

Separate within-group regressions—a series of regressions, one per group, of one or more dependent variables on a predictor variable. Compare **Pooled among-group regression** and **Pooled within-group regression**.

Shape—the geometry of the organism after information about scale, position, and orientation has been removed. Shape is not measureable. See **Form-change**.

Shape acceleration—increase of the ratio of **proportion-change** to **scale-change** in **form-comparisons** having the same **principal axes** with respect to some reference form.

Shape-change—see **Form-change**, **Proportion-change**.

Shape-change axis—in a scatter of logarithms of **principal dilatations**, the line $\log d_1 + \log d_2 = 0$. Compare **Size-change axis**.

Shape-complexity—see **Proportion-complexity**.

Shape coordinates for a triangle of **landmarks**—Cartesian coordinates for the location of the third landmark when the positions of the other two landmarks are arbitrarily fixed.

Shape deceleration—decrease of the ratio of **proportion-change** to **scale-change** in comparisons having the same **principal axes** with respect to some reference form.

Shape displacement—in a scatter of logarithms of **principal dilatations**, the perpendicular distance (always positive) of a point from the **size-change axis**; a measure of local **proportion-change**.

Shape isomery—invariance of the ratio of **proportion-change** to **scale-change** in **form-comparisons** having the same **principal axes** with respect to some reference form.

Shape variable—a product $X_1^{a_1} \ldots X_n^{a_n}$ of powers of measured distances for which the sum Σa_i of the exponents is 0.

Shear—the computation of a **sheared component**.

Sheared component, sheared factor—a second or subsequent **principal component** from which the effect of **general size** (or other appropriate **latent variable**) has been removed by subtracting a multiple of the first principal component according to an appropriate regression.

Singularity—a point at which a **deformation** is **isotropic**, representing a change in scale without change in proportions. At a singularity, **principal directions** for a **biorthogonal grid** cannot be computed.

Size—see **General size**, **Size measure**.

Size-change—see **Scale-change**.

Size-change axis—in a scatter of the logarithms of the **principal dilatations**, the line $\log d_1 = \log d_2$. Compare **Shape-change axis**.

Size-complexity—see **Scale-complexity**.

Size displacement—in a scatter of logarithms of **principal dilatations**, the perpendicular distance (positive or negative) of a point from the **shape axis**; a measure of local **scale-change**.

Size-free discriminator—a **linear combination** correlated with a group variable and uncorrelated, on the average, with **general size** within groups.

Size measure—any variable with an adequately large **loading** on **general size**.

Size variable—a product $X_1^{a_1} \ldots X_n^{a_n}$ of powers of measured distances for which the sum Σa_i of the exponents is 1.

Skeleton—see **Medial axis**.

Smooth—an attribute of a **deformation**: the quality of being spatially uninterrupted—having points nearby in one form corresponding to points nearby in the second form, and with straight line segments in one form not manifesting sharp angles in the second.

Standard size—an arbitrarily chosen value of a **size measure** used as the score on **general** or **primary size** in order to jointly predict values of **distance** measures.

Strain—a summary index of the extent to which a geometrical model, such as a set of points in a plane, fails to fit a collection of **distances** measured among landmarks.

Strain ratio—see **Dilatation**.

Symmetric axis—see **Medial axis**.

Tensor field (symmetric second-order tensor field)—the coordinate-free numerical representation of a **deformation**: a mathematical rule for consistently and smoothly assigning a **dilatation** in every direction through every point of a region. A tensor field is diagrammed using a little cross at every point of one of the forms. See **Principal axes**, **Principal dilatations**, **Biorthogonal grids**.

Transformation—see **Deformation**.

Trend surface—the modeling of a numerical function of position in a plane by a formula in terms of the coordinates (usually Cartesian) of position.

Triangulation—reconstruction of the relative positions of points by means of the measured distances among them taken three at a time. After an initial three points are placed, each subsequent point is located with respect to a pair located previously.

Triple point—center of a circle touching an **empirical boundary** at three points of tangency; intersection of three branches of a **medial axis**.

Truss—see **Box truss**.

Truss element—any **distance** measure between two **landmarks** of a **box truss**.

Unique variance—in **path analysis**, the additive component of a variable's observed value that has zero covariance with all other variables, latent or observed, in the model; also, the variance of this component. Compare **Communality**.

GLOSSARY

Unobserved variable—see **Latent variable**.

Variable—a quantitative **character**: any numerical quantity whose value is measured or inferred for each of a population of organisms. In this book, variables will ordinarily be **distances**.

Width—at a point upon the **medial axis**, the distance to the two or more nearest points of the **empirical boundary**.

Width function—a representation of a bilaterally symmetric **empirical boundary**: perpendicular distance to the axis of symmetry as a function of position along that axis.

Within-group size—the **general size factor** for a single group.

Literature Cited

Alberch, P., S. J. Gould, G. F. Oster, and D. B. Wake. 1979. Size and shape in ontogeny and phylogeny. Paleobiology 5(3):296-317.

Anderson, T. W. 1963. Asymptotic theory for principal component analysis. Annals of Mathematical Statistics 34(1):122-148.

Anstey, R. L., and D. A. Delmet. 1973. Fourier analysis of zooecial shapes in fossil tubular bryozoans. Geological Society of America Bulletin 84(5):1753-1764.

Atchley, W. R. 1971. A comparative study of the cause and significance of morphological variation in adults and pupae of *Culicoides*: a factor analysis and multiple regression study. Evolution 25(3):563-583.

Atchley, W. R., C. T. Gaskins, and D. Anderson. 1976. Statistical properties of ratios. I. Empirical results. Systematic Zoology 25(2):137-148.

Benson, R. H., R. E. Chapman, and A. F. Siegel. 1982. On the measurement of morphology and its change. Paleobiology 8(4):328-339.

Blum, H. 1973. Biological shape and visual science (Part I). Journal of Theoretical Biology 38(2):205-287.

Blum, H., and R. N. Nagel. 1978. Shape description using weighted symmetric axis features. Pattern Recognition 10:167-180.

Bolin, R. L. 1947. The evolution of the marine Cottidae of California with a discussion of the genus as a systematic category. Stanford Ichthyological Bulletin 3(3):153-168.

Bookstein, F. L. 1978a. The measurement of biological shape and shape change. Lecture Notes in Biomathematics, Vol. 24. Springer-Verlag, New York. 191 pp.

Bookstein, F. L. 1978b. Linear machinery for morphological distortion. Computers and Biomedical Research 11:435-458.

Bookstein, F. L. 1979. The line-skeleton. Computer Graphics and Image Processing 11:123-137.

Bookstein, F. L. 1980a. When one form is between two others: an application of biorthogonal analysis. American Zoologist 20(4):627-641.

Bookstein, F. L. 1980b. Data analysis by partial least-squares. Pp. 75-90 in: J. Kmenta and J. B. Ramsey (eds.), Evaluation of Econometric Models. Academic Press, New York. 410 pp.

Bookstein, F. L. 1981a. Looking at mandibular growth: some new geometric methods. Pp. 83-103 in: D. S. Carlson (ed.), Craniofacial Biology. University of Michigan Center for Human Growth and Development, Ann Arbor.

Bookstein, F. L. 1981b. Coordinate systems and morphogenesis. Pp. 265-282 in: T. G. Connelly, L. Brinkley, and B. Carlson (eds.), Morphogenesis and Pattern Formation. Raven Press, New York.

Bookstein, F. L. 1981c. Comment on "Issues related to the prediction of craniofacial growth." American Journal of Orthodontics 79:442-448.

Bookstein, F. L. 1982a. Foundations of morphometrics. Annual Review of Ecology and Systematics 13:451-470.

Bookstein, F. L. 1982b. The geometric meaning of soft modeling, with some generalizations. Pp. 55-74 in: K. G. Jöreskog and H. Wold (eds.), Systems under Indirect Observation: Causality, Structure, Prediction, Pt. 2. Contributions to Economic Analysis 139(1):1-292, (2):1-343. North-Holland Publ. Co., Amsterdam.

Bookstein, F. L. 1982c. On the cephalometrics of skeletal change. American Journal of Orthodontics 82(3):177-198.

Bookstein, F. L. 1983. The geometry of craniofacial growth invariants. American Journal of Orthodontics 83(3):221-234.

Bookstein, F. L. 1984a. A statistical method for biological shape change. Journal of Theoretical Biology 107:475-520.

Bookstein, F. L. 1984b. Tensor biometrics for changes in cranial shape. Annals of Human Biology 11:413-437.

Bookstein, F. L. 1985. Transformations of quadrilaterals, tensor fields, and morphogenesis. In: P. L. Antonelli (ed.), Mathematical Essays on Growth and the Emergence of Form. University of Alberta Press, Edmonton, Alberta, to appear.

Bookstein, F. L., P. D. Gingerich, and A. G. Kluge. 1978. Hierarchical linear modeling of the tempo and mode of evolution. Paleobiology 4(2):120-134.

Bookstein, F. L., R. E. Strauss, J. M. Humphries, B. Chernoff, R. L. Elder, and G. R. Smith. 1982. A comment upon the uses of Fourier methods in systematics. Systematic Zoology 31(1):85-92.

Burnaby, T. P. 1966. Growth-invariant discriminant functions and generalized distances. Biometrics 22:96-110.

Bussing, W. A. 1978. Taxonomic status of the atherinid fish genus *Melaniris* in lower Central America, with the description of three new species. Revista de Biologia Tropical 26(2):391-413.

Chernoff, B., J. V. Conner, and C. F. Bryan. 1981. Systematics of the *Menidia beryllina* complex (Pisces: Atherinidae) from the Gulf of Mexico and its tributaries. Copeia 1981(2):319-336.

Cherry, L. M., S. M. Case, J. G. Kunkel, J. S. Wyles, and A. C. Wilson. 1983. Body shape metrics and organismal evolution. Evolution 36(5):914-933.

Cheverud, J. M. 1982. Phenotypic, genetic, and environmental morphological integration in the cranium. Evolution 36(3):499-516.

Cheverud, J. M., J. L. Lewis, W. Bachrach, and W. D. Lew. 1983. The measurement of form and variation in form: an application of three-dimensional quantitative morphology by finite-element methods. American Journal of Physical Anthropology 62(2):151-165.

Dempster, A. P. 1969. Elements of Continuous Multivariate Analysis. Addison-Wesley, Menlo Park, California. 388 pp.

Douglas, M. E., and J. C. Avise. 1982. Speciation rates and morphological divergence in fishes: tests of gradual versus rectangular modes of evolutionary change. Evolution 36(2):224-232.

Ehrlich, R., R. B. Pharr, Jr., and N. Healy-Williams. 1983. Comments on the validity of Fourier descriptors in systematics: a reply to Bookstein et al. Systematic Zoology 32(2):202-206.

Eldredge, N., and J. Cracraft. 1980. Phylogenetic Patterns and the Evolutionary Process. Columbia University Press, New York. 349 pp.

Farris, J. S. 1979. The information content of the phylogenetic system. Systematic Zoology 28(4):483-519.

Fink, W. L. 1982. The conceptual relationship between ontogeny and phylogeny. Paleobiology 8(3):254-264.

Fornell, C. (ed.) 1982. A Second Generation of Multivariate Analysis, Vol. 1, Methods. Praeger, New York. 392 pp.

Fricke, R. 1982. Modification and use of McCune's shape measurement system for recent benthic fishes (Pisces). Braunschw. Naturk. Schr. 1(3):533-559.

Gans, C. 1960. Studies on amphisbaenids (Amphisbaenia, Reptilia). I. A taxonomic revision of the Trogonophinae and a functional interpretation of the amphisbaenid adaptive pattern. Bulletin of the American Museum of Natural History 119:129-204.

Gnanadesikan, R. 1977. Methods for Statistical Data Analysis of Multivariate Observations. John Wiley and Sons, New York. 311 pp.

Gould, S. J. 1966. Allometry and size in ontogeny and phylogeny. Biological Reviews of the Cambridge Philosophical Society 41(4):587-640.

Gould, S. J. 1977. Ontogeny and Phylogeny. Harvard University Press, Cambridge, Massachusetts. 501 pp.

Gould, S. J., D. S. Woodruff, and J. P. Martin. 1974. Genetics and morphometrics of *Cerion* at Pongo Carpet: A new systematic approach to this enigmatic land snail. Systematic Zoology 23(4):518-535.

Gregory, W. K. 1928. The body-forms of fishes and their inscribed rectilinear lines. Palaeobiologica 1(8):93-100.

Gregory, W. K. 1933. Fish skulls: a study of the evolution of natural mechanisms. Transactions of the American Philosophical Society 23(2):75-481.

Guttman, L. 1953. Image theory for the structure of quantitative variates. Psychometrika 18(4):277-296.

Hansell, R. I. C., F. L. Bookstein, and H. J. Rowell. 1980. Operational point homology by Cartesian transformation to standard shape: examples from setal positions in phytoseiid mites. Systematic Zoology 29(1):43-49.

Hennig, W. 1966. Phylogenetic Systematics. University of Illinois Press, Urbana, Illinois. 263 pp.

Hilbert, D., and S. Cohn-Vossen. 1952. Geometry and the Imagination. Chelsea Publishing Co., New York. 357 pp.

Hopkins, J. W. 1966. Some considerations in multivariate allometry. Biometrics 22:747-760.

Hubbs, C. L. 1941. The relation of hydrological conditions to speciation in fishes. Pp. 182-195 in: A Symposium on Hydrobiology. University of Wisconsin Press, Madison.

Humphries, J. M., F. L. Bookstein, B. Chernoff, G. R. Smith, R. L. Elder, and S. G. Poss. 1981. Multivariate discrimination by shape in relation to size. Systematic Zoology 30(3):291-308.

Huxley, J. S. 1932. Problems of Relative Growth. The Dial Press, New York. 276 pp.

Jardine, N. 1967. The concept of homology in biology. British Journal of Philosophical Science 18:125-139.

Jardine, N. 1969. The observational and theoretical components of homology: a study based on the morphology of the dermal skull-roofs of rhipidistian fishes. Biological Journal of the Linnean Society of London 1:327-361.

Jolicoeur, P. 1963. The multivariate generalization of the allometry equation. Biometrics 19:497-499.

Jolicoeur, P., and J. E. Mosimann. 1960. Size and shape variation in the painted turtle, a principal component analysis. Growth 24(4):339-354.

Jöreskog, K. G., and H. Wold (eds.). 1982. Systems under Indirect Observation: Causality, Structure, Prediction. Contributions to Economic Analysis 139(1):1-292, (2):1-343. North-Holland Publ. Co., Amsterdam.

Kluge, A.G., and W. C. Kerfoot. 1973. The predictability and regularity of character divergence. American Naturalist 107(955):426-442.

Liem, K. F., and L. S. Kaufman. 1984. Intraspecific macroevolution: functional biology of the polymorphic cichlid species *Cichlasoma minckleyi*. Pp. 203-215 in: A. A. Echelle and I. Kornfield (eds.), Evolution of Fish Species Flocks. University of Maine at Orono Press, Orono, Maine. 257 pp.

Lohmann, G. P. 1983. Eigenshape analysis of microfossils: a general morphometric procedure for describing changes in shape. Mathematical Geology 15(6):659-672.

Lull, R. S., and S. W. Gray. 1949. Growth patterns in the Ceratopsia. American Journal of Science 247(7):492-503.

Maynard Smith, J. 1968. Mathematical Ideas in Biology. Cambridge University Press, London. 152 pp.

McCune, A. R. 1981. Quantitative description of body form in fishes: implications for species level taxonomy and ecological inference. Copeia 1981(4):897-901.

Miller, R. R., and A. Carr. 1974. Systematics and distribution of some freshwater fishes from Honduras and Nicaragua. Copeia 1974(1):120-125.

Mosimann, J. E. 1970. Size allometry: size and shape variables with characterizations of the log-normal and generalized gamma distributions. Journal of the American Statistical Association 65(330):930-945.

Mosimann, J. E. 1975. Statistical problems of size and shape. I. Biological applications and basic theorems. Pp. 187-217 in: G. P. Patil, S. Kotz, and K. Ord (eds.), Statistical Distributions in Scientific Work, Vol. II. D. Reidel Publ. Co., Dordrecht, Holland.

Mosimann, J. E., and F. C. James. 1979. New statistical methods for allometry with application to Florida red-winged blackbirds. Evolution 33(1):444-459.

Moss, M. L., R. Skalak, G. Dasgupta, and H. Vilmann. 1980. Space, time, and spacetime in craniofacial growth. American Journal of Orthodontics 77:591-612.

Moss, M. L., R. Skalak, M. Shinozuka, H. Patel, L. Moss-Salentijn, H. Vilmann, and P. Mehta. 1983. Statistical testing of an allometric centered model of craniofacial growth. American Journal of Orthodontics 83(1):5-18.

Moyers, R. E., and F. L. Bookstein. 1979. The inappropriateness of conventional cephalometrics. American Journal of Orthodontics 75:599-617.

Olson, E. C., and R. L. Miller. 1958. Morphological Integration. University of Chicago Press, Chicago. 317 pp.

Oxnard, C. E. 1980. The analysis of form: without measurement and without computers. American Zoologist 20(4):695-705.

Pimentel, R. A. 1979. Morphometrics. Kendall/Hunt, Dubuque, Iowa.

Potter, F. E., Jr., and S. S. Sweet. 1981. Generic boundaries in Texas cave salamanders, and a redescription of *Typhlomolge robusta* (Amphibia: Plethodontidae). Copeia 1981(1):64-75.

Rao, C. R. 1973. Linear Statistical Inference and its Applications, 2nd ed. John Wiley and Sons, New York. 625 pp.

Raup, D. M. 1966. Geometric analysis of shell coiling: general problems. Journal of Paleontology 40(5):1178-1190.

Richards, O. W., and A. J. Kavanagh. 1943. The analysis of the relative growth gradients and changing form of growing organisms: illustrated by the tobacco leaf. American Naturalist 77(772):385-399.

Richards, O. W., and A. J. Kavanagh. 1945. The analysis of growing form. Pp. 188-230 in: W. E. Le Gros Clark and P. B. Medawar (eds.), Essays on Growth and Form presented to D'Arcy Wentworth Thompson. Clarendon Press, Oxford. 408 pp.

Rohlf, F. J., and J. W. Archie. 1978. Least-squares mapping using interpoint distances. Ecology 59(1):126-132.

Rohlf, F. J., A. J. Gilmartin, and G. Hart. 1983. The Kluge-Kerfoot phenomenon—a statistical artifact. Evolution 37(1):180-202.

Rosen, D. E., and R. M. Bailey. 1963. The poeciliid fishes (Cyprinodontiformes), their structure, zoogeography, and systematics. Bulletin of the American Museum of Natural History 126:1-176.

Salmon, G. 1914. A treatise on the analytic geometry of three dimensions, 2 vols. (R. A. P. Rogers, rev. ed.). Longmans and Green, London. 470 and 334 pp.

Schopf, T. J. M. 1982. A critical assessment of punctuated equilibria. I. Duration of taxa. Evolution 36(6):1144-1157.

Serra, J. P. 1982. Image Analysis and Mathematical Morphology. Academic Press, New York. 610 pp.

Siegel, A. F., and R. H. Benson. 1982. A robust comparison of biological shapes. Biometrics 38(2):341-350.

Skalak, R., G. Dasgupta, M. Moss, E. Otten, P. Dullemeijer, and H. Vilmann. 1982. Analytical description of growth. Journal of Theoretical Biology 94:555-577.

Smith, R. J. 1980. Rethinking allometry. Journal of Theoretical Biology 87:97-111.

Sneath, P. H. A. 1967. Trend-surface analysis of transformation grids. Journal of Zoology 151:65-122.

Sokal, R. R., and F. J. Rohlf. 1981. Biometry, 2nd ed. W. H. Freeman and Company, San Francisco. 859 pp.

Sprent, P. 1972. The mathematics of size and shape. Biometrics 28(1):23-37.

Stanley, S. M. 1979. Macroevolution. W. H. Freeman and Company, San Francisco. 332 pp.

Stoermer, E. F., and T. B. Ladewski. 1982. Quantitative analysis of shape variation in type and modern populations of *Gomphoneis herculeana*. Beiheft zur Nova Hedwigia 73:347-386.

Strauss, R. E. 1980. Genetic and morphometric variation and the systematic relationships of eastern North American sculpins (Pisces: Cottidae). Dissertation, Pennsylvania State University, University Park.

Strauss, R. E., and F. L. Bookstein. 1982. The truss: body form reconstructions in morphometrics. Systematic Zoology 31(2):113-135.

Stroud, C. P. 1953. An application of factor analysis to the systematics of *Kalotermes*. Systematic Zoology 2(2):76-92.

Taliev, D. N. 1955. Sculpins of Baikal (Cottoidei). Academy of Sciences U.S.S.R., Eastern Siberian Branch, Baikal Limnological Station, Publication House of the Academy of Sciences, U.S.S.R., Moscow. 603 pp.

Tattersall, I. 1973. Cranial anatomy of the Archaeolemurinae (Lemuroidea, Primates). Anthropological Papers of the American Museum of Natural History 52(1):1-110.

Thompson, D'A. W. 1961[1917,1942]. On Growth and Form, abridged edition, J. T. Bonner (ed.). Cambridge: The University Press. 346 pp.

Thorpe, R. S. 1976. Biometric analysis of geographic variation and racial affinities. Biological Reviews of the Cambridge Philosophical Society 51(4):407-452.

Thorpe, R. S. 1980. A comparative study of ordination techniques in numerical taxonomy in relation to racial variation in the ringed snake *Natrix natrix* (L.). Biological Journal of the Linnaean Society 13:7-40.

Tobler, W. R. 1977. Bidimensional regression: a computer program. W. R. Tobler, Santa Barbara, California. 72 pp.

Tobler, W. R. 1978. The comparison of plane forms. Geographical Analysis 10:154-162.

Tukey, J. W. 1954. Causation, regression, and path analysis. Pp. 35-66 in: O. Kempthorne, T. A. Bancroft, J. W. Gowen, and J. L. Lush (eds.), Statistics and Mathematics in Biology. Iowa State College Press, Ames, Iowa. 632 pp.

Tyler, J. C. 1980. Osteology, phylogeny, and higher classification of the fishes of the Order Plectognathi (Tetraodontiformes). National Oceanic and Atmospheric Administration Technical Report, National Marine Fisheries Service Circular 434:1-422.

Van Valen, L. 1965. The study of morphological integration. Evolution 19(3):347-349.

Van Valen, L. 1974. Multivariate structural statistics in natural history. Journal of Theoretical Biology 45:235-247.

Webber, R. L., and H. Blum. 1979. Angular invariants in developing human mandibles. Science 206(4419):689-691.

Wigner, E. P. 1967. The unreasonable effectiveness of mathematics in the natural sciences. Pp. 222-237 in: Symmetries and Reflections: Scientific Essays of Eugene P. Wigner. University of Indiana Press, Bloomington, Indiana; M.I.T. Press, Cambridge, Massachusetts.

Woodger, J. H. 1945. On biological transformation. Pp. 95-120 in: W. E. Le Gros Clark and P. B. Medawar (eds.), Essays on Growth and Form presented to D'Arcy Wentworth Thompson. Clarendon Press, Oxford. 408 pp.

Wright, S. 1932. General, group and special size factors. Genetics 17(5):603-619.

Wright, S. 1954. The interpretation of multivariate systems. Pp. 11-33 in: O. Kempthorne, T. A. Bancroft, J. W. Gowen, and J. L. Lush (eds.), Statistics and Mathematics in Biology. Iowa State College Press, Ames, Iowa. 632 pp.

Wright, S. 1968. Evolution and the Genetics of Populations. Vol. I: Genetic and Biometric Foundations. University of Chicago Press, Chicago. 469 pp.

Younker, J. L., and R. Ehrlich. 1977. Fourier biometrics: harmonic amplitudes as multivariate shape descriptors. Systematic Zoology 26(3):336-342.

Index

Acceleration, 98, 199–202
    see also Shape acceleration
Affine transformation, 11, 20–21, 22n, 56
Alberch, P., et al., 198–201
Allometric coherence, 192–198
Allometric factor model, 25, 36
Allometric reassociation, 192–198
Allometry
    and averaging of forms, 124–126
    characterization of, 24, 142–161
    and dilatations, 145, 148–152
    group differences in, 157–158
    in the first principal component, 86
    and interspecific divergence, 153–158
    models for, 11, 24–25, 35
    and morphospace, 207
    nonlinearities in, 98, 195
    and orthogonal decompositions, 40
    and ratios, 226
    and size-standardization, 124–126
    statistical analysis of, 237–238
Analytic boundary, 64–67, 132
Angles, use of, 46, 72, 74–80
    formulas for, 215–217
Archive, morphometric, 8, 37, 46, 52–54, 69
    alternate versions of, 144–146
Area, as measure of size, 163n, 217, 238
Atherinella
    growth of, 95–100, 148–152
    landmarks for, 144–145
Atherinidae
    digitizing of, 66–67
    examples involving, 144–152
Averaged forms, 124–126, 156–157

Belonesox, example involving, 138–144
Biological homology, *see* Homology
Biorthogonal analysis, 14–23, 127–142
    computation of, for triangles, 218–225, 239–242
    examples of, 132–142, 148–150, 156–157
    and evolutionary inference, 135–138
    and factor analysis, 144–158
    of factors, 197–198
    and heterochrony, 200–205
    interpolation underlying, 190–192
    inversion of, 135–137
    parameters of, 162–163
    of reassociation, 197–198
    selection of forms for, 146, 197–198
    and shear analysis, 146–149, 156–157
    of size-standardized forms, 141, 145, 156–157, 165
    steps of, 130–131, 134, 137
    of triangles, 218–225
    see also Complexity, Dilatations
Biorthogonal directions, 16–23
    see also Biorthogonal analysis
Biorthogonal grids, *see* Biorthogonal analysis
Bisector, angle, formula for, 217
Bisector, perpendicular, formula for, 216
Blum, H., 70–73
Body size, *see* Standard body size, General size
Boundary curve, 8, 136, 140
    digitizing of, 63–67
Boundary homology function, 64, 132
Box truss, *see* Truss network

Canonical analysis, *see* Discriminant analysis
Cartesian coordinates, 50–52
    formulas involving, 215–217, 231
    logarithms of dilatations as, 162–167
Cartesian grid, 4–5, 14–16, 128–129, 140–142
Catostomidae, examples involving, 168–186
Character, Character set, 9, 36, 187–188
    from complexity analysis, 179
    and morphospace, 206–207
    selection of, 46, 192, 206–207

Circle
    as boundary arc, 64–67, 132, 141
    as coordinate curve, 226–228
    as descriptor of shape change, 23
Cladistics, 1, 179, 209–213
Coherence, allometric, 192–198
Communality, 88–89
Complexity, 162, 166–186
Component, *see* Principal components, Shear analysis
Computed-homology, Computed-homologues, 6, 190–192; *see also* Deformation model
Conic arc
    in analytic boundary, 64
    as coordinate curve, 226–227
Consistency criterion
    in path analysis, 81, 86, 89
    in shear analysis, 102–105
Convergence, 152, 182–186
Coordinates, Coordinate systems, 8, 50–52, 225–229
    and measurement of shape change, 11, 14–23; *see also* Biorthogonal analysis
    role in choosing landmarks, 6–8
    for single forms, unimportance of, 14
    multivariate analysis of, 230–238
Coplanarity criterion, 54
Correlation matrix, 26
    in factor analysis, 85, 87–89, 92–93
    and morphological integration, 194–195
Cosine, formulas for, 215–216
Cottidae, example involving, 208–210
Cottus
    averaged forms for, 125–126
    examples involving, 153–161
    secondary factors in, 94–98
    truss network for, 55, 57–61
Covariance matrix, 26
    adjusted for secondary factors, 122–124
    adjusted for size, 102–104
    in allometric coherence and reassociation, 193–198
    see also Path analysis
Covariance structure, analysis of, *see* Path analysis

Cursor, 61–62

Data, morphometric, 45–67
    correct form of, 23–26
    and meristic, 115–117
    three-dimensional, 189–191
        see also Archive, Coordinate, Landmarks
Deformation model for homology, 4–11, 23, 127
    biorthogonal depiction of, 127–131
    limits of, 135–136, 140–141, 172
    over small regions, 16–17
        see also Biorthogonal analysis
Developmental program, 198–205
Diatoms, example using, 40–42
Digitizing, 61–67
    noise of, 140–142
Digitizing tablet, 50, 61–62
Dilatations, 14, 16–19, 21–22, 130–131
    computation of, for triangles, 239–242, 240–244
    gradients of, 134–135, 193
    and heterochrony, 200–205
    logarithms of, as Cartesian coordinates, 162–167
    regional analysis of, 162–168
    after size-standardization, 165–166
    variance of, *see* Complexity
        see also Biorthogonal analysis
Diodon, example involving, 132–139
Direct effect, in path analysis, 80
Discriminant analysis, Discrimination
    allometry and, 142–161
    in *Cottus*, 159–161
    by features of the medial axis, 73–80
    Fisherian, as path analysis, 78, 82–84
    by residuals, 110–115
    and secondary factors, 95, 110–115
    and size difference, *see* Shear analysis
    suboptimal, 37–43
Displacement, 163–168
Dissimilarity, taxonomic, 206–207
Dissociation, 196
    regional, 162
Distance, formulas for, 215–216
"Distance," from shape difference measures,

179, 184–185
Distance measures
  alignment of, 45
  and biorthogonal analysis, 127, 149, 152
  conventional, 45–46
    for *Atherinella*, 146
    compared with biorthogonal grid, 149
    for *Cottus*, 155, 160
    for *Gila*, 120
    compared with truss network, 159–161
  and coordinates, 225–228
    Cartesian, 50–52
  coverage of form, 45–46, 48, 60
  diagonal, 146, 149–152, 159–161; *see also* Truss network
  and homology, 152
  loadings of, 46, 48, 147–161, 193
  log transforms of, 23–26
  patterns of, 45–57, 127
  and principal directions, 152
  redundancy of, 45–46, 52–54
  at standard body size, 124
  truss elements as, 53
Divergence
  and allometry, 105, 153–158
  ontogeny and phylogeny of, 183–186
  and taxonomic dissimilarity, 206–207
Dual space, 28

Ecophenotypes, in *Cottus*, 153–161
Eigenvectors, 86, 89, 92–93; *see also* Principal components
Ellipse
  as boundary arc, 64–65
  as coordinate curve, 226–227
  as descriptor of shape change, 21–22, 48
Empirical boundary, 64
End point, of medial axis, 70, 74, 79
Evolutionary biology in the morphometric context, 1, 187–213
Evolutionary inference
  from biorthogonal grids, 135–138
  from complexity analysis, 175–186
  and developmental programs, 198–205

Factor analysis, Factor model, 9–11, 24–25, 46, 85–102
  and biorthogonal analysis, 144–158, 197–198
  computational considerations, 89–98, 245–250
  confirmatory, 93
  and correlations, 86–89
  of landmark position, 94–100
  for nonlinearities of allometry, 98, 195
  path diagrams for, 87, 90, 95, 100
  compared with principal components, 86, 89–94
  secondary, *see* Secondary factors
  for three variables, 87–89
Fink, W., 198–201
Fins, digitizing of, 66–67
Flattening a tetrahedron, 54–56
Foraminifera, example using, 40–42
Foreshortening, 98
Form represented as a mathematical function, 37–40
Form-change, 10–11, 14, 142–161
Form-comparisons, 1–238
Forms
  averaged, 124–126
  unobserved, 206–212
Fourier analysis, 38–43, 212
Function, as representation of form, 37–40

Gambusia, example involving, 138–144
General size, 9, 48, 85–89, 105, 149–152, 190, 193n
Geographic variation, 153–161
Geometric form, 6–7, 63–65, 212; *see also* Archive
Gila, example involving, 115–122
Gomphoneis, example involving, 40–42
Gould, S., 192, 196, 198–201
Gradients, of dilatation, 134–135, 193–198
Grid, *see* Biorthogonal analysis, Cartesian grid
Groups, delineation of, 107
Growth, 73–80, 142–161
  accretionary, 188–190

complexity of, 175–176
and phyletic comparisons, 175–186
simulation of, 124–126
Growth gradient, 193–198

Helping point, 64, 132, 141
Heterochrony, 192, 198–205
Heterogeneity of shape change, 161–186, 198
Homologues, anatomical and material, 189–191; *see also* Homology function
Homology, biological, 3–8, 23
in relation to geometric form, 37–43, 60, 67
of meristic features, 189–190
Homology function, Homology map, 4–8, 10, 14–15, 188–192, 212
on the boundary, 64
and the Cartesian grid, 4–5, 128–129, 140, 142
and digitization, 65–67
as an interpolation, 130–131
suggested by medial axes, 72–73
see also Homology, biological; Archive
Homoscedasticity, 26
Hopkins, J., 195, 204
Hotelling's T^2 statistic, 234–237
Huxley, J., 24–25, 193–194
Hybrids, 115–121
Hyperbola
as boundary arc, 64–65
as coordinate curve, 226–227
Hypermorphosis, 122, 199–202
Hyperplane, 28

Indirect effect, in path analysis, 80
Inflection point, 63–65
Intermediacy, 115–117
Inversion of form-comparisons, 128, 130, 135–137, 162–165
Isoclines of strain, 208–212
Isometry, 24–25, 236
Isotropic transformation, 23

Kluge-Kerfoot phenomenon, 35–36

Landmarks, 5–14
anatomical, 46
for biorthogonal analysis, 131–132
and digitizing, 64–67
and distance measures, 45–46
duplication of, 188–190
examples of
for atherinids, 66–67
for catostomids, 170
for poeciliids, 138–141
for tetraodontiformes, 133
extremal, 46
factors for position of, 94–100
internal, 191–192
from medial axis, 73–79
in relation to orthogonal decompositions, 38, 41–43
selection of, 190–191
source
photographs, 170
radiographs, 145
statistical analysis of, 230–238; *see also* Factor analysis
see also Archive; Distance measures; Homology map
Latent variable, *see* Factor model
Legendre polynomials, 38, 40–42
Leghorn chickens, example involving, 89–93, 195
Lines, formulas involving, 215–217
coordinates for, 228–229
Line-element, 16, 20–22
Loading, 11, 46, 48, 102, 127, 157, 193; *see also* Distance measures
Log shape variables
defined, 29
discarded, 35
Log size variables
defined, 28
discarded, 35
Log-transformation, 23–26
Longitudinal data, 85, 144, 195

Macroevolution, 197
Mandibles, human, example using, 72–74

Index

Mean forms, *see* Averaged forms
Measurement, *see also* Distance measures
   precision of, 52–54, 60, 62
   unit of, 25, 36
Medial axis, 70–80
Melaniris, see *Atherinella*
Meristic features, homology of, 189–190
Meristic variables, 102, 115–121
Mesh points, 130, 132, 146, 166–168; *see also* Biorthogonal analysis
Micropterus, example involving, 73–80
Models, morphometric, 2–11
   multivariate, 78–101
      see also Factor model; Deformation model
Mola, example involving, 132–139
Monophyletic groups, 208–212
Morphological integration, 194–198
Morphological space, Morphospace, 206–212
Morphometric models, *see* Models
Mosimann, J., 25, 30–32
Multiple regression, 78–81
Multivariate analysis of landmark coordinates, 230–238

Neoteny, 199–202
Net shape displacement, 166–168
Net size displacement, 166–168
Normal equations, in path analysis, 79–81
Numerical taxonomy, 1

Olson, E., and R. Miller, 194–195
Ontogeny
   and phyletic comparisons, 175–186, 199–200
   time scale for, 199–205
      see also Growth
Opercles, measurement of, 73–80
Orientation, effect of, 3, 50, 74; *see also* Distance measures
Orthogonal components, 38–40, 86
Orthogonal decomposition, 38–40
Orthogonal functions, 38–40
Outgroup comparisons, 175–186, 210–212
Outline, *see* Boundary curve

Paedomorphosis, 177, 199–202
Parabola
   as boundary arc, 64–67
   as coordinate curve, 226–227
Path analysis, 78–101
   and allometric coherence, 192–195
   and reassociation, 196–198
Path coefficient, *see* Path analysis, Path diagram
Path diagrams, 79–101, 111, 194
Peramorphosis, 199–202
Perpendicularity, formulas for, 216–217; *see also* Cartesian coordinates
Phenetics, 1
Phylogeny, and ontogenetic comparisons, 175–186, 199–200
Pimentel, D., 90
Poeciliidae, example involving, 138–144
Point mode, of a digitizing tablet, 61–62
Point of inflection, 63, 65
Points, formulas involving, 215–217
Polar coordinates, 225–227
Post-displacement, 199–202,
Pre-displacement, 199–202
Precision, of landmark locations, 52–54
Primary size, 92–93, 114, 193–198
Principal axes, 16–22, 130–131; *see also* Biorthogonal analysis
Principal components
   compared with biorthogonal analysis, 148–152
   as eigenvectors, 26, 86
   compared with factors, 86, 89–94
   in the study of morphospace, 208–212
   of an orthogonal decomposition, 40–42
   orthogonality of, 86
   path analysis of, 78, 86–87
   regression of size out of, 102–109, 153
   residuals from, 33–35
   of residuals, 110–115
   shape and size confounded in, 102–103, 115, 117
   shear analysis of, 102–110
Principal dilatation, *see* Dilatations
Principal directions, *see* Biorthogonal analysis

Progenesis, 199–202
Proportion-change, 161–186
Proportion-complexity, 167–168, 172
Pseudolandmark, 67, 141

Quadrilateral, 53–57
Quadruple point, of medial axis, 74, 78

Radius function, 37–40
Ranzania, example involving, 135–139
Ratios, 3, 27–32, 46; *see also* Dilatations
Reassociation, allometric, 192–198
Reflection, of point in line, formula for, 217
Registration, choice of, 3, 50, 74
Regression
    models for, 33
    as path analysis, 78–81
    for size-standardization, 124–126
        see also Residuals
Residuals
    from correlations, 87, 92–98
    as shape variables, 33–35
    from size, 84, 103–105, 114n, 153
        see also Regression
Resistant fit, 11, 56
Retardation, 98
ρF-sets, 194–195
Rohlf, F., and J. Archie, 52–53
Rotation, of factors, 86

Scale-change, 161–186
Scale-complexity, 167–168, 172
Scorpaena, example involving, 210–212
Secondary factors, 92–100
    and allometric coherence, 193–198
    in biorthogonal analysis, 198
    in discrimination, 95, 110–115
    removal of, by shearing, 122–124
Sequestering of correlations, 92–93
Shape, 1–238
Shape acceleration, 202–205
Shape-change, 142–186, 198–205
    rudiments of, 14–22
Shape-change axis, 163–166
Shape coordinates, 230–238
Shape deceleration, 202–205
Shape difference
    characters from, 179
    as component, 103–105
    as "distance," 179, 184–185
    measures of, 161–175
    and size-standardization, 156–157
    and taxonomic dissimilarity, 206–207
Shape displacement, 163
Shape isomery, 202–205
Shape variables, 27–35, 200–205
Shear analysis, 101–109
    and biorthogonal analysis, 144–158
    examples of, 146–149, 153–161
    in identification of hybrids, 115–122
    as regression, 84, 103–104, 123
    SAS routine for, 243–245
    and secondary factors, 110–115, 122–124
Shearing (geometric), 142, 149–152
Significance, tests of statistical, 213, 230–238
Simulation of growth, 124–126
Sine, formulas for, 216
Singularity, 23, 167n
Size, 11, 24–25, 36, 85–87, 101–103
    and discrimination, 82–84, 110–115
    residuals from, correlation of, 114n
    and shape
        confounded in first two principal components, 102–103, 115, 117
        connection between, 28–30
        correlations between, 30–32, 238
        difference between, 102–105
    in shear analysis, 103–105
    and taxonomic dissimilarity, 207
Size adjustment, 26–36
Size-change axis, 163–166
Size displacement, 163–167
Size factor, *see* General size, Primary size, Secondary Factors
Size-free shape discrimination, 82–84; *see also* Shear analysis
Size standardization, 124–126, 141, 145, 153, 156–157, 165–167
Size variables, 28–35, 124, 153, 238
    see also General size
Skeleton, *see* Medial axis
Skulls, fish, examples involving, 138–142, 168–186

Index

Smooth mapping, 6, 127, 189–191; *see also* Deformation model
Sneath, P., 128
Stage, ontogenetic, 199–205
Standard body size, 61, 124–126, 150, 156
Standard deviation of morphometric variables, 26
Standard length, and secondary factors, 95–100
Statistical significance, tests of, 213, 230–238
Strain, as measure of measurement error, 57–62
  and morphospace, 208–212
  and size-standardization, 125–126
Strain ratio, *see* Dilatation
Stream mode, of a digitizing tablet, 62–64
Symmetric axis, 70–80
Systematics, 1, 210–213

T^2 statistic, 234–237
Tangent line, 64–67
Tensor field, 10, 130–131; *see also* Biorthogonal analysis
Tent diagram, 184–185
Tetrahedron, 54–57
Tetraodontiformes, example involving, 132–139
Thompson, D'A., 4–5, 14–16, 128–130, 187
Three-dimensional data, 2, 131n, 189–191
Time scale, ontogenetic, 200–205
  phylogenetic, 206
Transformation
  isotropic, 23
  logarithmic, 23–26
  power, 25–26
  see also Deformation model

Transformation series, 179
Trend surface, 128
Triangles, 20–23, 52–54
  biorthogonal analysis of, 218–225
  statistical analysis of, 230–238
Triangulation, 46–49, 52–53
Triple point, of medial axis, 70–71, 74, 79
Truss network, 52–61, 124–126
  and biorthogonal analysis, 156–157
  compared with conventional distance measures, 159–161
  for *Cottus*, 57–61, 155–161
  and growth simulation, 125–126
  and secondary factors, 94–98
  size measures from, 61, 124
  and unobserved forms, 206–212

Unique variance, 110, 121, 193–196
Unit of measurement, 25, 36

Variables, 8–11, 23–36
  correlations among, 194–195
  homology and, 37–43, 51–52
  unobserved, *see* Factor model
Variance
  between and within groups, 35–36
  total, 197
  unique, 121, 193–196
  see also Covariance matrix

Width, medial, 70, 72
Width function, 38, 41
Within-group size, 83–84, 103, 105; *see also* General size
Wright, S., 78–101, 195